De Gruyter Textbook

Krischer, Schönleber • Physics of Energy Conversion

Also of Interest

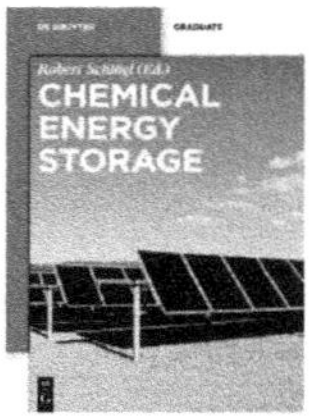

Chemical Energy Storage
Robert Schlögl (Ed), 2012
ISBN 978-3-11-026407-4, e-ISBN 978-3-11-026632-0

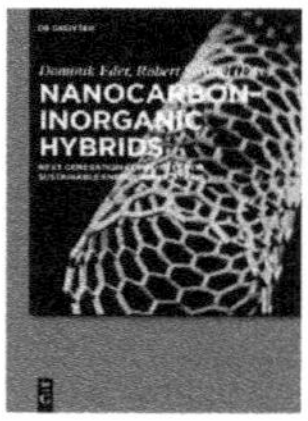

Nanocarbon Inorganic Hybrids: Next Generation Composites for Sustainable Energy Applications
Dominik Eder, Robert Schlögl, 2014
ISBN 978-3-11-026971-0, e-ISBN 978-3-11-026986-4

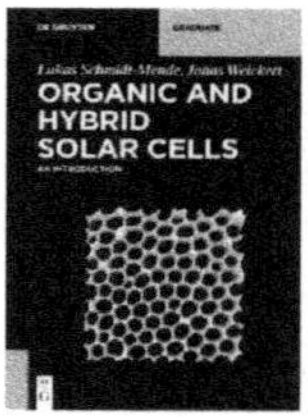

Organic and Hybrid Solar Cells: An Introduction
Lukas Schmidt-Mende, Jonas Weickert, 2016
ISBN 978-3-11-028318-1, e-ISBN 978-3-11-028320-4

Wind Energy Harvesting: Micro-to-Small Scale Turbines
Ravi Kishore, Colin Steward, Shashank Priya, 2015
ISBN 978-1-61451-565-4, e-ISBN 978-1-61451-417-6

Energy Harvesting and Systems
Shashank Priya (Editor-in-Chief)
ISSN 2329-8774, e-ISSN 2329-8766

www.degruyter.com

Katharina Krischer and Konrad Schönleber

Physics of Energy Conversion

DE GRUYTER

ISBN 978-1-5015-0763-2
e-ISBN (PDF) 978-1-5015-1063-2
e-ISBN (EPUB) 978-1-5015-0268-2

Library of Congress Cataloging-in-Publication Data
A CIP catalog record for this book has been applied for at the Library of Congress.

Bibliographic information published by the Deutsche Nationalbibliothek
The Deutsche Nationalbibliothek lists this publication in the Deutsche Nationalbibliografie; detailed bibliographic data are available on the Internet at http://www.dnb.de.

Typesetting: PTP-Berlin, Protago TEX-Production GmbH
Printing and binding: CPI books GmbH, Leck
Cover image: LOWELL GEORGIA/Science Source/getty images
♾ Printed on acid-free paper
Printed in Germany

www.degruyter.com

Preface

Among the current central challenges of mankind are problems related to a sustainable energy supply, the growing global energy demand, and the climate change due to an increasing level of CO_2 in the atmosphere. Discussions on such energy-related questions have reached the public at the latest since the beginning of the 21st century, and articles on energy and climate topics are omnipresent in the news. However, for laypersons it is often difficult to draw correct conclusions from the information or to assess it critically. Terms are frequently used in an imprecise way or not defined, different aspects of a problem are mixed, or only part of the picture is given. For an open societal discourse, better education of the public in general is necessary, and it is the task of universities to contribute to this.

Students of the natural sciences, in particular physicists and chemists, and of engineering should thus be provided with a good understanding of energy sciences, i.e. with the physical foundations of energy conversion processes. Such a basis is necessary for a competent dialog, and has to be incorporated in other aspects of the energy discussion, such as economic, social or political points of view. Yet, even when isolating the scientific side of this transdisciplinary topic, a lecturer faces the problem that the topic seems to be too broad to be covered within one lecture in a consistent way. For example, power plants, usually taught in engineering, should be covered as well as solar cells, a topic native to semiconductor physics or fuel cells, which are typically taught in classes of electrochemistry or physical chemistry.

It is the aim of this textbook to provide a unified view on the different energy conversion processes. The approach is driven by thermodynamics: each energy conversion device can be described as an open thermodynamic system, and efficiencies for the ideal converter as well as the realistic converter can be assigned. In this context, it is necessary to be aware of the actual work into which a certain amount of energy input can be converted. This becomes most apparent when trying to interpret the colloquial expression "energy consumption", which contradicts the principle of conservation of energy and thus also the first law of thermodynamics. It turns out that the maximal amount of work that can be extracted from a given amount of energy is a good measure for its value. This physical quantity is called exergy or availability, with the first term being adopted in this book. The exergy quantifies the quality of a certain amount of input energy, and it is this property which is, in fact, referred to when talking about "energy consumption". Thus, the concept of exergy allows for the comparison of the quality of different forms of energy such as heat, chemical energy, or electrical energy.

Accordingly, the central theme is the treatment of energy converters as open systems and the performance of efficiency analyses, based on the concept of exergy. This naturally leads to the chosen outline of the book: The introduction of exergy in closed systems in Chapter 2, followed by an introduction to open systems and the transfer of the exergy concept to this class of systems in Chapter 3. Equipped with this knowledge, we proceed to discuss the two most important heat engines, steam power plants, and gas turbine power plants in Chapter 4. This discussion starts with the thermodynamically ideal, i.e. reversible, processes, and continues with some aspects of the design criteria of real power plants. However, here as well as in the following chapters, we concentrate on the fundamental physical description rather than the concrete implementation and state of the art realizations. Hence, for example, materials science aspects are to the largest part left aside.

Chapters 5 and 6 treat the thermodynamics of electricity and chemical reactions, respectively. The former connects solid state physics and thermodynamics, the latter summarizes chemical thermodynamics. These concepts are then used in Chapter 7, which is devoted to electrochemical energy conversion. As electrochemistry is often not taught in the regular curriculum, the chapter starts with an introduction to important electrochemical concepts before again evaluating ideal and realistic conditions of conversion of chemical into electrical energy. Again, it is our goal to elaborate on a unified thermodynamic description of galvanic cells, which should bring the student in a position to understand more specialized works on fuel cells or batteries, but the different realizations of these classes of devices are not the topic of this book.

Chapters 8, 9, and 10 are then devoted to solar energy. The properties of solar radiation, including its thermodynamic description, are discussed in Chapter 8. Chapters 9 and 10 are then devoted to the two common solar energy conversion routes, namely solarthermal and photovoltaic energy conversion. As before, the thermodynamic principles operative in these devices when viewed as open systems are the thread of the argumentation. In contrast, different types of realizations, such as the diverse types of solar cells and their material properties play a minor role. Especially in Chapter 10 the unifying description of the energy conversion process becomes apparent. The physics of solar cells combines concepts developed in Chapters 3, 5, 7, and 9. The final chapter, Chapter 11, does not develop novel concepts, but briefly summarizes relevant possibilities for exergy storage, applying the principles developed throughout the book.

This book is written for students in physics, chemistry, engineering, or related disciplines in their last year of bachelor studies, but it might also serve as a textbook for an introductory master level course on energy sciences, which would then be ideally complemented by one or more advanced courses on state of the art devices and current research and development problems in a particular type of system, such as solar cells, fuel cells, or power plants. It is expected that the students have a basic knowledge of phenomenological thermodynamics of closed systems and of solid state physics. The fundamentals of these two subjects are summarized in Appendices A and B, respectively. Depending on the actual curriculum (or disciplines) of the students in the course, it might be necessary to review these topics within the lecture to a greater or lesser extent.

The book is based on a lecture given several times for physics students in their last year of bachelor studies at the Technische Universität München (TUM). We had several discussions with the tutors of the accompanying exercises and several other colleagues, and we would like to thank them for their input. In particular, we thank Qi Li, Simon Filser, and Katrin Bickel for their critical reading of the manuscript, helpful comments, and advice. Special thanks also to Helen Shiells for her linguistic advice and proof reading of the manuscript. Furthermore, we acknowledge funding by TUM.solar in the framework of the Bavarian Collaborative Research Project "Solar technologies go hybrid" (SolTec).

Munich, April 2015 Katharina Krischer and Konrad Schönleber

Contents

Preface — v

1 Introduction — 1
1.1 Terms and definitions — 1
1.1.1 Units — 5
1.1.2 Example: Energy consumption and production — 6
1.2 Energy conversion processes — 7
1.2.1 Exergy — 11
1.2.2 Conversion efficiency — 12
1.2.3 Example: heating of a room — 13

2 Exergy in closed systems — 15
2.1 Thermodynamic basics — 15
2.1.1 The first law of thermodynamics — 16
2.1.2 The second law of thermodynamics — 17
2.2 Exergy in closed systems — 18
2.3 Exergy transfers — 21
2.3.1 Exergy transfer via work transfer — 22
2.3.2 Exergy transfer via heat transfer at $T = T_\mathrm{h}$ — 22
2.3.3 Exergy and anergy contents of transferred energy — 24
2.3.4 The laws of thermodynamics in terms of exergy — 24
2.4 Exergy sources — 25
2.5 Heat pumps — 26
2.5.1 Working principle — 27

3 Exergy in open systems — 31
3.1 Thermodynamics of open systems — 31
3.1.1 The first law of thermodynamics for open systems — 32
3.1.2 Steady flow conditions — 34
3.1.3 Technical work — 35
3.2 Exergy of open systems — 37
3.3 Important example components — 38
3.3.1 Heat exchangers — 38
3.3.2 Water turbines — 41
3.3.3 Wind turbines — 41

4 Thermal power plants — 45
4.1 Steam power plants — 45
4.1.1 Rankine cycle — 46

4.1.2 Efficiency of a steam power plant — 51
4.1.3 Example calculation — 55
4.1.4 Modifications in a real steam power plant — 57
4.2 Gas turbine power plants — 60
4.2.1 Joule–Brayton cycle — 62
4.2.2 Optimization criteria — 64
4.2.3 Efficiency of a gas turbine power plant — 66
4.2.4 Example calculation — 67
4.2.5 Intercooling — 69
4.2.6 Combined cycle power plants (CCPP) — 71

5 Electrical exergy — 73
5.1 The electrochemical potential — 73
5.1.1 Interfaces — 75
5.1.2 Electrical currents — 77
5.2 Voltage sources — 78
5.3 Generators — 79
5.3.1 Electrical power output — 80
5.3.2 Mechanical power input — 81
5.4 Thermoelectrics — 82
5.4.1 Seebeck coefficients — 83
5.4.2 Thermoelectric energy conversion — 84

6 Chemical exergy — 87
6.1 Basic concepts — 87
6.1.1 Example calculation — 89
6.2 The driving force of a chemical reaction — 90
6.2.1 Chemical activity — 91
6.2.2 The driving force of a chemical reaction at a given state — 94
6.3 The exergy of fuels — 94
6.4 Efficiency of the combustion process — 97

7 Electrochemical energy conversion — 99
7.1 Electrochemistry — 99
7.1.1 The standard hydrogen scale — 100
7.1.2 Origin of the electrode potential — 101
7.1.3 Electrode potential and cell voltage — 103
7.1.4 The Nernst equation — 105
7.1.5 Electrochemical voltage sources — 106
7.2 Electrochemical energy conversion — 108
7.2.1 Maximal efficiency — 108
7.2.2 Efficiency of a realistic galvanic cell — 110

7.2.3 Reaction kinetics — 112
7.2.4 Fuel cells — 115
7.2.5 Nonrechargeable batteries — 116

8 Solar energy — 119
8.1 Properties of solar irradiation — 119
8.1.1 Stefan–Boltzmann law — 122
8.1.2 Exergy of the solar irradiation — 122
8.2 Influence of the atmosphere — 124
8.2.1 The greenhouse effect — 126
8.3 Photosynthesis — 127
8.3.1 Conversion efficiency — 129

9 Solarthermal energy conversion — 131
9.1 Heat radiation of the absorber — 133
9.1.1 Selective absorption — 135
9.2 Concentrating devices — 136
9.2.1 Imaging optics — 136
9.2.2 Technical realization — 139
9.2.3 Efficiency of concentrating solarthermal energy converters — 140
9.2.4 Solarthermal power plants — 141

10 Photovoltaic energy conversion — 143
10.1 Conversion of solar into chemical energy — 145
10.1.1 Chemical energy and exergy at a given n_e and n_h — 145
10.1.2 Radiative recombination — 149
10.1.3 Maximal efficiency of a solar cell — 152
10.2 Selective contacts — 155
10.3 Conversion of chemical into electrical energy — 157
10.3.1 Current voltage characteristic — 158
10.3.2 Electrical power output — 160
10.3.3 Losses — 161
10.4 Solar cell design — 163
10.4.1 c-Si solar cell — 163
10.4.2 Other types of solar cells — 164

11 Exergy storage — 167
11.1 Mechanical exergy storage — 167
11.1.1 Pumped storage hydro power plant — 168
11.1.2 Compressed air energy storage — 168
11.2 Thermal exergy storage — 169
11.2.1 High temperature storage — 170

11.2.2 Low-temperature storage — 171
11.3 (Electro-)Chemical exergy storage — 172
11.3.1 Rechargeable batteries (Li-ion batteries) — 174
11.3.2 Water splitting — 175
11.3.3 Power to gas — 176

A Basics: Thermodynamics — 177
A.1 Thermodynamic potentials — 177
A.2 Ideal gases — 178
A.2.1 Heat capacity — 178
A.2.2 State changes in ideal gases — 179
A.3 The Carnot cycle — 182
A.4 Phase transitions — 183

B Basics: Solid state physics — 187
B.1 Particle ensembles — 187
B.1.1 Density of states — 187
B.1.2 Distribution functions and the electrochemical potential — 188
B.1.3 Selected properties — 189
B.1.4 The Sommerfeld expansion — 191
B.2 Semiconductors — 192
B.2.1 Intrinsic semiconductors — 192
B.2.2 Doping — 194
B.2.3 Optical properties — 194

C Further Reading — 197

Index — 199

1 Introduction

It is a central challenge for humanity to convert its energy infrastructure into a sustainable one. Sustainable energy supply means that the energy needs of the present are met without restricting the ability of future generations to satisfy their needs, as defined in the 'Agenda 21' by the United Nations. Only such an energy infrastructure is able to sustain a high standard of living in the future. To achieve this goal the sources of energy used should at least satisfy the following conditions:

1. they should not be consumed;
2. they should not lead to the emission of environmental pollutants;
3. they should not lead to health risks and inherent social injustices.

All three conditions are violated by the most common energy sources used today, e.g. fossil fuels. While the first condition can be strictly met by the renewable energy sources such as water, wind, and solar power, the other two can only be approximated even in these cases. For example, water power requires huge dams with possible severe consequences for the people and animals living in that area. In order to minimize the potential harm while maximizing the benefit, it is hence necessary to make a careful evaluation of the specific energy needs and the possible mix of sources. This evaluation is a complex process and requires a detailed understanding of the energy supply technologies and the processes involved. It is the goal of this book to provide such a detailed understanding on the physical level for the important processes involved in the generation of electrical energy.

1.1 Terms and definitions

In the discussion of energy consumption, three terms with different meanings are frequently used: **primary energy**, **final energy**, and **net energy**. These terms are also synonymously applied to powers instead of energies, i.e. **primary power**, **final power**, and **net power**.

Primary energy

The total amount of energy stored in the different natural energy sources, such as fossil fuels, is called primary energy. Primary energy cannot be used directly but has to be refined in a process that takes away a part of the energy. The end product of the refinery process is the final energy. Refinery first means the refinery process necessary to produce energy in the form of directly usable fuels, such as diesel or gasoline, from natural sources such as crude oil. More importantly, another process that can also be seen as a refining of energy is the conversion of primary energy carriers, for example coal, gas, or uranium, into electrical energy as final energy. It is the latter process that accounts for most of the difference between primary and final energy.

For **fossil fuels** the primary energy is the **heating value** (old: lower heating value), i.e. the total amount of heat released if the products of the combustion of coal, oil, or natural gas are all in the gaseous state. This value differs from the **upper heating value**, which is the heat of combustion obtained when all combustion reactants and products are at a standard temperature of $T = 25\ ^{\circ}C$. In the latter case, water is in the liquid state, the difference between heating value and upper heating value being the **heat of vaporization** or **latent heat** of water. The upper heating value is therefore always larger than the heating value.

The other sources of primary energy such as **nuclear fuels** and **renewable energies** are treated according to the following standard method:

The electricity produced is divided by an assumed efficiency for the conversion of the natural energy source into electrical energy. This assumed efficiency is 33 % for nuclear power and 100 % for renewable energies. The optimal conversion efficiency for renewable energies is justified by the idea that the actual amount of energy from sunlight or wind is infinite and cannot be consumed, making the produced electrical energy the energy source. With this method the amount of primary energy stemming from the conventional sources fossil fuels and nuclear power is overestimated compared to renewable energies.

Another no longer official method for calculating the primary energy of these nonfossil energy sources is the substitution method, where the primary energy is associated with the amount of fossil fuels saved by the use of nonfossil fuel energy sources. With this method the share of renewable energies has a higher value; for example, 1 kWh of renewable electricity can replace ca. 2.5 kWh of coal bound primary energy and is thus counted as 2.5 kWh primary energy.

Final energy

Final energy is the energy that is readily delivered to the consumers. Examples are electrical energy or the heating values of refined organic fuels. The share of renewable energies in the final energy is higher than the share in the primary energy, as the conversion efficiency of conventional power plants and refinery processes is already factored in. For example, 1 kWh of coal bound primary energy will produce no more than 0.4 kWh of final energy in the form of electricity, whereas primary and final energies are the same for renewable energy sources. Typically the overall final energy for all energy sectors is about 2/3 of the overall primary energy as long as efficient energy converting systems are used.

Net energy

The net energy is the energy in its final desired form, i.e. the energy output of the consuming device. The conversion efficiency associated with the consumer is responsible for the difference between final and net energy. The overall net energy output is about 1/2 of the final energy consumption, leading to a rough estimate of the ratios of primary to final to net energy of 3 : 2 : 1. Already these very basic considerations show, that the most efficient way to reduce the consumption of primary energy is to reduce the amount of net energy consumption. This can be achieved for example by minimizing transport routes, heating rooms to lower temperatures in winter, or decreasing consumption of nonessential goods and services. Any unit net energy saved roughly translates into three units of primary energy saved.

Two examples will illustrate the differences between the different energy terms:

1. Room illumination:
In this process the purpose of the lamp is to provide a certain number of photons in the visible range per unit of time. The energy of these photons is hence the net energy. The lamp itself is driven by electrical energy, which is the final energy in this case. Hence, the efficiency of the lamp gives the conversion factor between final and net energy. The electricity is provided by a power plant or a renewable energy source. In the former case the primary energy needed is either the heating value of the required fossil fuel (e.g. coal) plus the energy taken out during the refinery of this fuel, or three times the final energy if a nuclear power plant is providing the energy. If a renewable energy source is used, the final and primary energies are identical.

2. Driving a car:
Here the net power is the mechanical power of the engine. The final energy in this case is the heating value of the fuel, e.g. gasoline, which differs from the primary energy only by the energy taken away by the refinery process. In the case of an electric car the final energy is electrical energy and thus identical to the final energy in the first example, and everything with regards to primary energy can be taken from there.

All three quantities specifying energy consumption are important and are unfortunately often confused in public discussions. Care has to be taken especially looking at the share of renewable energies in the overall energy mix, as this might have political implications. As an example, in Figure 1.1 the respective shares of different sources of primary energy are shown together with the share of renewable energies in the final energy for both the European Union (EU 28) and the USA in 2012. The sources are public data given out by the relevant statistics agencies EUROSTAT and the Energy Information Administration of the Department of Energy for the EU and the USA, respectively.

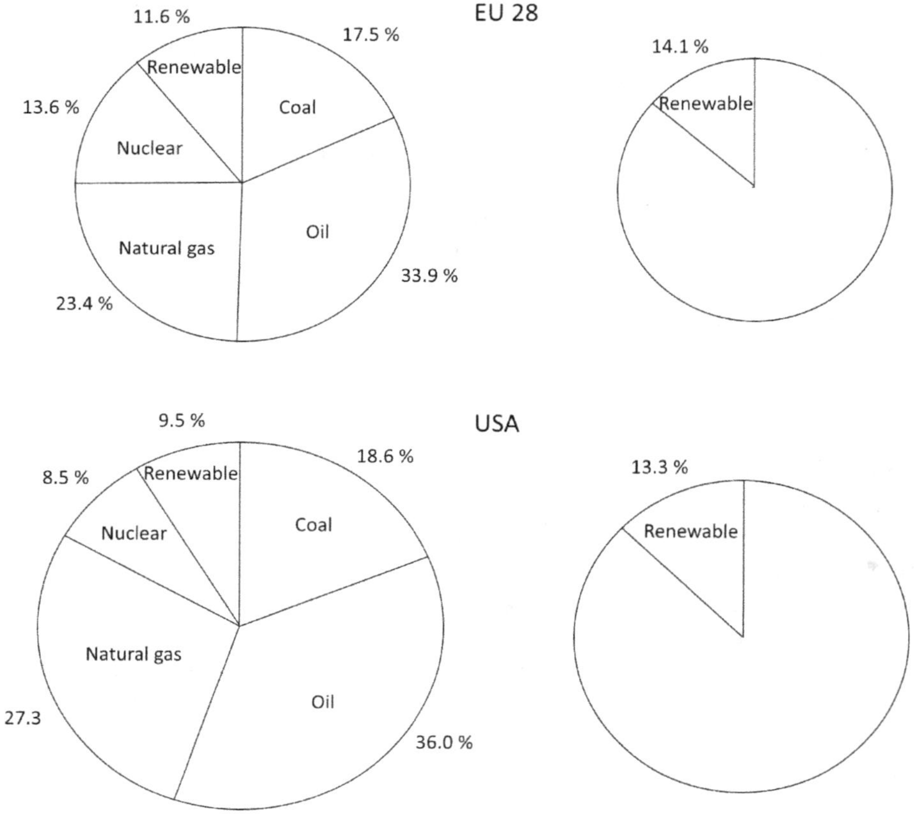

Fig. 1.1: *Left:* Respective shares of the different sources of primary energy for both the EU (top) and the USA (bottom) in the year 2012. *Right:* contribution of renewable energies to the mix of final energy of the EU (*top*) and the USA (*bottom*). The total areas of the circles are proportional to the total amounts of energy consumed (*left:* EU 1.68 Gtoe, USA 2.44 Gtoe; *right*: EU 1.10 Gtoe, USA 1.74 Gtoe). Data are taken from EUROSTAT (http://epp.eurostat.ec.europa.eu/portal/page/portal/energy/data/main_tables) and the DOE (U.S. Energy Information Administration, Monthly Energy Review, August 2014) for the EU and the USA, respectively.

As evident from Figure 1.1 the primary energy consumed in both the EU and the USA stems mostly from the fossil fuels coal, oil, and natural gas. In the EU the share of these energy carriers is close to 75 %, while it exceeds 80 % in the USA. Furthermore, in both regions the shares of renewable energies in primary and final energy, respectively, are substantially different due to the calculation method of primary energy discussed above.

As primary, final, and net energy are typically not all of the same basic form, conversion processes between the forms of energy with a certain efficiency are necessary. For

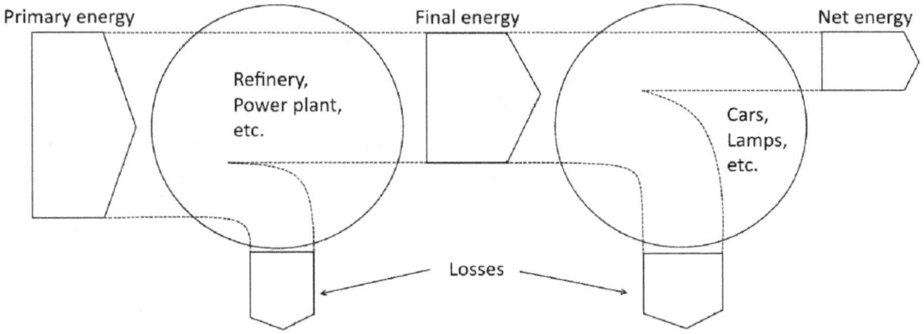

Fig. 1.2: Energy flow through a converter from primary to final energy (*left circle*) and a converter from final to net energy (*right circle*). The widths of the arrows are proportional representations of the amounts of energy in the respective forms.

a visualization of the energy conversion processes, so-called **energy flow diagrams** as shown in Figure 1.2 are often used. In these diagrams the losses in the individual converters are the differences of the widths of the arrows, leaving the converters on the right-hand side and the widths of the arrows entering the converters on the left-hand side, i.e. they are equal to the width of the arrows pointing downward.

Of course, the conversion processes represented by the two circles in Figure 1.2 can be broken down further. Examples of more detailed energy flow diagrams will be shown for the different energy conversion processes discussed throughout this book. As a rule of thumb, the losses in the first conversion step from primary to final energy are relatively small for the production of fossil fuels, and about 2/3 of the total primary energy input for the production of electricity. This does not mean that the overall losses in the conversion from primary to net energy are always higher when the intermediate form of final energy is a chemical fuel. The second conversion process from final to net energy generally shows substantially lower losses when electrical energy is the intermediate form of final energy. The reason for this is the higher value of electrical energy as opposed to high temperature heat as provided by a chemical fuel that is burnt. It is the evaluation of this quality difference between different forms of energy that will form the basis of the discussions in the rest of this book.

1.1.1 Units

The typical units used are the **watt**, *W*, for the power and the **kilowatt hour**, *kWh*, for the energy. Other units for energy frequently used in different contexts are: **joule**, *J*; **electron volt**, *eV*; **calorie**, *cal*; **tonne of coal equivalent**, *tce*; and **tonne of oil equivalent**, *toe*. A conversion table is given in Table 1.1. The joule J is the SI unit (international system of units) for energy and thus widely in use. It is the amount of energy

Table 1.1: Conversion table for the different energy units frequently used. Another unit sometimes used in English-speaking countries is the *British thermal unit* or *btu*. It has a value of 1100 J.

	kWh	J	eV	cal	tce	toe
1 kWh =	1	$3.6 \cdot 10^{6}$	$2.3 \cdot 10^{25}$	$8.6 \cdot 10^{5}$	$1.2 \cdot 10^{-4}$	$8.6 \cdot 10^{-5}$
1 J =	$2.8 \cdot 10^{-7}$	1	$6.3 \cdot 10^{18}$	0.24	$3.4 \cdot 10^{-11}$	$2.4 \cdot 10^{-11}$
1 eV =	$4.4 \cdot 10^{-26}$	$1.6 \cdot 10^{-19}$	1	$3.8 \cdot 10^{-20}$	$5.4 \cdot 10^{-30}$	$3.8 \cdot 10^{-30}$
1 cal =	$1.2 \cdot 10^{-6}$	4.2	$2.6 \cdot 10^{19}$	1	$1.4 \cdot 10^{-10}$	$1.0 \cdot 10^{-10}$
1 tce =	8100	$2.9 \cdot 10^{10}$	$1.8 \cdot 10^{29}$	$7.0 \cdot 10^{9}$	1	0.7
1 toe =	$1.2 \cdot 10^{4}$	$4.2 \cdot 10^{10}$	$2.6 \cdot 10^{29}$	$1.0 \cdot 10^{10}$	1.4	1

required to maintain a force of one newton over a distance of one meter. More importantly in the context of this book, it also marks the amount of energy transferred when one coulomb of charges is moved across an electrostatic potential difference of one volt. By this second equality the electron volt is linked to the joule by the elementary charge constant $e = 1.6 \cdot 10^{-19}$ C as it marks the energy transferred when one elementary charge is moved across an electrostatic potential difference of one volt. The units calorie and tons of coal and oil equivalents are all defined by the process of heating water. One calorie is the amount of energy needed to heat one gram of water from 14.5 °C to 15.5 °C at normal pressure. The ton of oil equivalent is then defined by convention as 10 Gcal, which is the amount of energy required to heat 10 000 tons of water from 14.5 °C to 15.5 °C at normal pressure. It amounts to the heat released when burning one ton of crude oil. Note that sometimes conventions for the exact value of 1 toe are used which are slightly different from the one given here. The ton of coal equivalent, in turn, is defined as 7 Gcal. In the context of this book mostly the joule is used with the exception of two important cases. First, when household or large-scale energy consumption is considered, the kWh is the most common unit and typically also the unit of energy price calculations. In the latter case, e.g. for statistics of the world energy consumption or the energy consumption of a specific domestic economy, tce or toe are also sometimes used. Second, when energy conversion processes are broken down to the energy transferred by single charge carriers, or in general in the quantum regime, by far the most convenient unit is the eV.

1.1.2 Example: Energy consumption and production

To get a feeling for the amount of energy needed for certain tasks and the powers involved, a few examples will be briefly discussed here. A human being has a thermal power of 80–100 W, leading to a daily net energy need of about 2000 kcal or ca. 2.3 kWh/d just to keep the body running and the body temperature constant. This value is of course increased by the work we do every day, with peak powers around 500 W. All these values are net energies. The final energy stored in the consumed food is higher, since the energy conversion efficiency of the body, and especially the efficiency of the digestive system, has to be taken into account. Note that the relevant primary energy consumption associated with food production is not only the renewable energy stored in the food itself but also the energy consumption for farming and transportation.

On the energy-supply side, typical output powers of common power plants are of interest. The amount of electrical final power provided by typical power plants are ca. 0.3–5 MW for wind turbines, 0.1–2 GW for coal fired, and 1–3 GW for nuclear power plants. Rooftop photovoltaic power converters have a typical peak power density of 1 kW_p/m^2, while the largest solar thermal power plants have peak powers of several 100 MW.

According to the International Energy Agency, the total primary energy consumption of humanity was ca. 13.4 Gtoe/a in 2012, which means that the 7.1 billion human beings then alive had an average primary energy consumption of ca. 1.9 toe/a (2012), corresponding to an average primary power consumption of ca. 2.6 kW per capita (based on IEA data from “Key World Energy Statistics 2014” © OECD/IEA 2014, IEA Publishing; modified by the authors Licence: http://www.iea.org/t&c/termsandconditions/). This number is considerably higher in industrialized countries, with values of ca. 4.6 kW (40 MWh/a) primary power consumption per capita in the European Union and 11 kW (94 MWh/a) in the USA (both 2012) (data are taken from EUROSTAT (http://epp.eurostat.ec.europa.eu/portal/page/portal/energy/data/main_tables) and the DOE (U.S. Energy Information Administration, Monthly Energy Review, August 2014) for the EU and the USA, respectively). For these two regions energy flow diagrams are shown in Figure 1.3.

Mostly the production of electrical energy is responsible for the conversion losses from primary into final energy. The energy productivity, i.e. the amount of GDP (gross domestic product) enabled by one toe primary energy, is $ 6400 in the USA and $ 9700 in the EU. As large parts of the world population are aiming at the standard of living of the industrialized countries, the total worldwide primary energy consumption is steeply increasing (> 100 % increase between 1973 and 2012). For this reason it is doubly important to find ways to maintain the advantages of the high standard of living of a modern industrialized country at a much lower level of primary energy consumption stemming from renewable sources.

1.2 Energy conversion processes

The major objective of this book is the evaluation of energy conversion processes, which naturally splits into two parts. First, the energy conversion pathway itself has to be considered, and, second, the efficiency of the devices used to realize this pathway. As a very important example, consider the conversion of thermal energy into mechan-

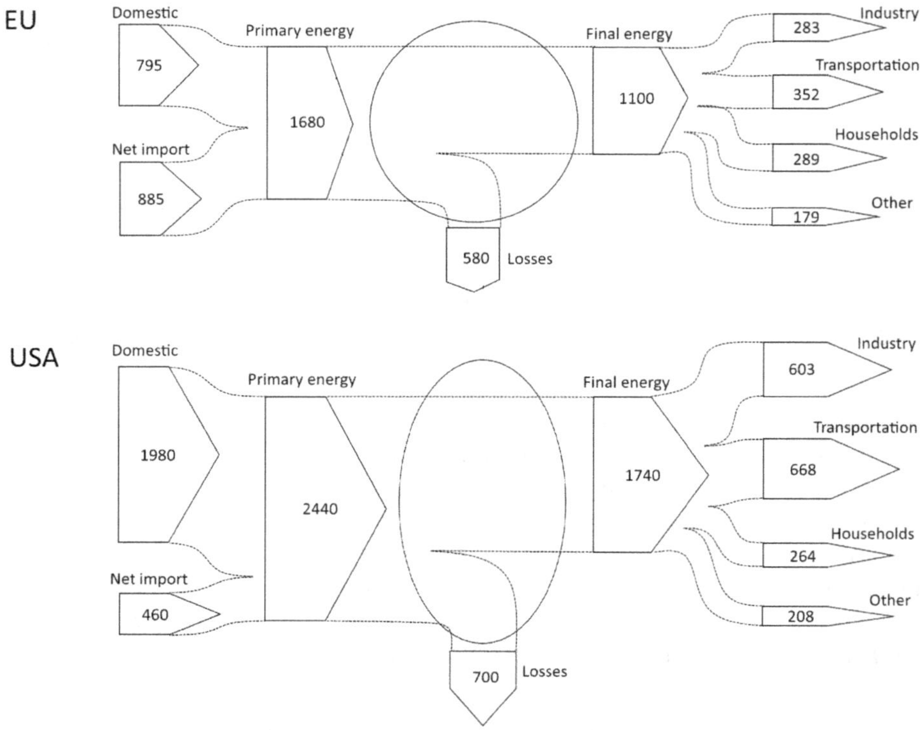

Fig. 1.3: Energy flow diagram for the 28 member states of the European Union combined (*top*) and the USA (*bottom*), both for the year 2012. The widths of the arrows are proportional representations of the amounts of energy in the respective forms, and all numbers given are in *Mtoe* (megatons oil equivalent). In addition to the energy conversion process from primary to final energy (*oval*), the sources of primary energy are broken down into domestic and imported values, and the final energy consumption is broken down by sector. Data are taken from EUROSTAT (http://epp.eurostat.ec.europa.eu/portal/page/portal/energy/data/main_tables) and the DOE (U.S. Energy Information Administration, Monthly Energy Review, August 2014) for the EU and the USA, respectively.

ical work. It is fair to say that exactly this question was the main driver that led to the development of the entire field of thermodynamics during the 19th century, a process not coincidentally accompanied by the increasing application of heat engines during the industrial revolution. The question of how much mechanical work can be gained by a periodically working heat engine was famously answered by Sadi Carnot in 1824 and later clarified and reformulated in modern terms by R. Clausius. This question and its solution is at the heart of the assessment of the quality of a given form of energy and is very instructive for all considerations made in this book. It will thus be discussed here in some detail and is also elaborated in Section A.3 of the appendix. In this discussion first an ideal, i.e. reversible process is assumed. In doing so the evaluation of the process itself is properly separated from the evaluation of a specific implementation of the process.

Consider a heat engine working periodically between two fixed temperature levels, $T_h > T_c$. An amount of heat Q provided isothermally at a temperature T in a reversible process is exactly given by $Q = \Delta ST$, with ΔS being the change in the entropy in the thermodynamic system which the heat enters. This increase of entropy in the system is transported by the heat itself, i.e. the heat input is accompanied by an entropy input of ΔS. The input heat Q_{in} at the higher temperature level of the heat engine T_h thus takes the value $Q_{in} = \Delta ST_h$. In the assumed reversible processes the entropy in the heat engine is conserved in all state changes not involving heat transfers. This means that the same amount of entropy that enters the heat engine has to be discarded in each cycle. To achieve this, an amount of heat $Q_{out} = \Delta ST_c$ has to be rejected at the lower temperature level. Thus, the maximal mechanical energy output $E_{mech,out}$ by a periodically working heat engine is given by

$$E_{mech,out} = Q_{in} - Q_{out} = \Delta S(T_h - T_c) \tag{1.1}$$

and the maximal efficiency η of the heat engine then correspondingly amounts to

$$\eta = \frac{E_{mech,out}}{Q_{in}} = \frac{\Delta S(T_h - T_c)}{\Delta ST_h} = 1 - \frac{T_c}{T_h} \tag{1.2}$$

It is the necessity to discard parts of the energy at the lower temperature level due to the entropy conservation in a reversible process that leads to the overall efficiency being lower than $\eta = 1$. This necessity arises as the heat engine is supposed to be working in a periodic fashion, and is thus not allowed to continuously accumulate entropy. Only if the lower temperature level approaches absolute zero temperature is $\eta = 1$ approached. The efficiency given in equation (1.2) is called the **Carnot efficiency** η_C and marks, as was shown, the maximal work extractable from a periodically driven heat engine working between the temperature levels T_h and T_c.

There is an analogy, which Carnot in fact considered, between the ideal reversible heat engine just described and a waterwheel. Though this analogy falls short in certain important aspects, it is still illustrative in others. For a waterwheel when an amount m of water is added at the inlet at a height of h_{in}, the energy input E_{in} is the potential energy in the gravitational field of the earth, i.e. $E_{in} = mgh_{in}$. The amount m of water then crosses the waterwheel and is discarded at the outlet at height h_{out}, a process associated with an energy loss of $E_{loss} = mgh_{out}$. In a perfect process, the maximal amount of energy extractable from the waterwheel E_{out} is thus given by $E_{out} = mg(h_{in} - h_{out})$, and the efficiency η of the process is consequently given by

$$\eta = \frac{E_{out}}{E_{in}} = \frac{mg(h_{in} - h_{out})}{mgh_{in}} = 1 - \frac{h_{out}}{h_{in}}. \tag{1.3}$$

Entropy in the case of the reversibly working the heat engine and mass in the case of the waterwheel cannot be destroyed in the respective energy converters, and their

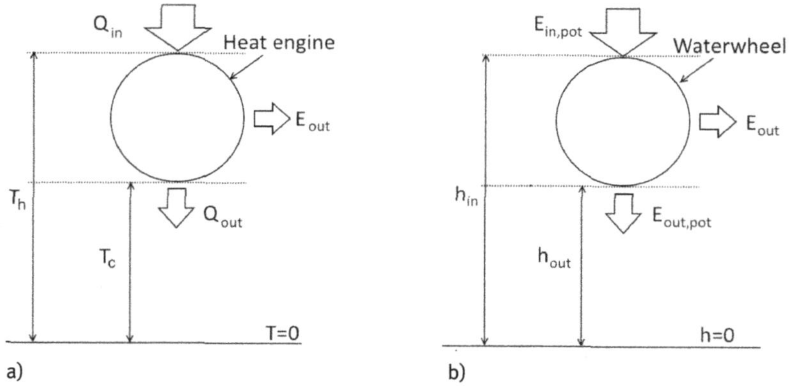

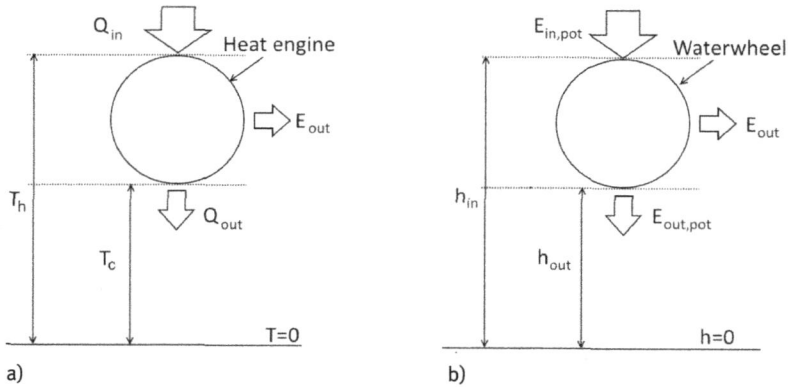

Fig. 1.4: Analogy between (a) a heat engine operating between two temperature levels and (b) a waterwheel operating between two height levels.

release is associated with an energy loss, due to the nonzero values of the lower temperature and height level, respectively. In both cases it is this necessity to discard a part of the input energy that leads to the nonunity value of the maximal efficiency. A diagrammatic visualization of both energy conversion processes is given in Figure 1.4.

Typically, one will not find the efficiency given in equation (1.3) in the context of waterwheels. The reason for this is that the energy input part, i.e. the transporting of water from sea level to the higher height level h_{in} by means of solar-driven evaporation, subsequent rainfall, and the formation of rivers, is not even considered. Instead of using the absolute altitude as a measure for the energy input, the river feeding the waterwheel, and thus a steady cost-free energy input, is taken as a given. It is also this line of thought that leads to the notion of the final energy output rather than the energy input being the primary energy delivered by renewable energies. The efficiency typically considered when waterwheels are discussed does not evaluate the chosen energy conversion path itself, as discussed here, but the quality of the device performing the energy conversion. In other words, it is concerned with the second part of the energy evaluation process discussed below. The efficiencies given in equations (1.2) and (1.3) mark the maximal possible conversion efficiencies for the processes chosen. An ideal process was assumed in both cases, and no amount of clever engineering could improve these values without changing the basic temperature or height parameters of the processes. It is therefore helpful to introduce the notion of a *value* inherent to different forms of energy and clever engineering is then indeed required to build machines that retain most of this value in real energy conversion processes. The definition of a quantity measuring the value of energy in different forms will be done in the following section.

1.2.1 Exergy

Strictly speaking, the term energy consumption, though frequently used, is a violation of the first law of thermodynamics (see equation (2.1) below), as energy can only be transformed from one form to another, but not consumed. Still, an amount of final energy provided by e.g. fuel or electricity can certainly not be used multiple times for the same purpose, e.g. moving a car or illuminating a room. Whenever the term energy consumption is used, the depletion of energy that can be used for certain purposes is meant. In the former example, by moving the car once, the ability of the energy initially stored in the fuel to move the car a second time is lost. The fuel is burnt and the combustion heat converted in a heat engine into the kinetic energy of the car and waste heat. Once the fuel is burnt, friction forces gradually convert the kinetic energy of the car into waste heat, leading to a conversion of the entire originally useful energy stored in the fuel into waste heat. No energy has been destroyed in the process, but the original fuel is gone and the energy stored therein is converted into low temperature heat which is no longer of any use to move a car. The same is true for the illumination of a room with an electrical lamp. The visible photons generated are partly absorbed at every surface they encounter and gradually converted to waste heat which can no longer be used to illuminate the room.

The key question to determine the value of energy sources, as just exemplified, is which part of their energy content can be converted into mechanical work or, as will be shown later, into electrical energy. It is this ability of the energy source to perform work that determines its value. The second law of thermodynamics states that this maximal extractable work in general is not identical with the amount of energy itself (for details see Section 2.1.2). The part of the energy that can be converted into useful work is called **exergy** or **availability**, Φ, the rest is called **anergy**, A. Any amount of energy, E, can thus be decomposed into these two contributions:

$$E = \Phi + A. \tag{1.4}$$

The term energy consumption in most cases actually means exergy consumption, as the ability of the energy to perform work is depleted while the total amount of energy remains constant.

The concept of exergy allows for the comparison of energies given in different forms such as electrical energy, heat provided at different temperatures, or energy stored in chemical bonds. The quantity *exergy* is only well defined when a reference state, the **ambient state**, is given. Usually this ambient state is defined as: $p = 1$ bar, $T = 25$ °C and concentrations of all chemicals as in the atmosphere, but especially the ambient temperature may vary somewhat around the standard temperature in different situations of interest, making another definition of the reference state more appropriate. In this book the ambient temperature is, however, set to the value of $T_{am} = 25$ °C for all calculations.

1.2.2 Conversion efficiency

For the conversion process of an amount of energy in one form, E_{in}, into energy in another desired form, E_{out}, an efficiency η_1 can be given:

$$\eta_1 = \frac{E_{out}}{E_{in}}. \quad (1.5)$$

This efficiency is called **energy or first-law efficiency**. It is the widely-used term for efficiency and was also used in equations (1.2) and (1.3). It has, however, only limited value when one tries to assess the design of a given energy conversion process itself. This can be visualized by so-called **exergy flow diagrams**, as shown in Figure 1.5, which are modified versions of energy flow diagrams discussed above.

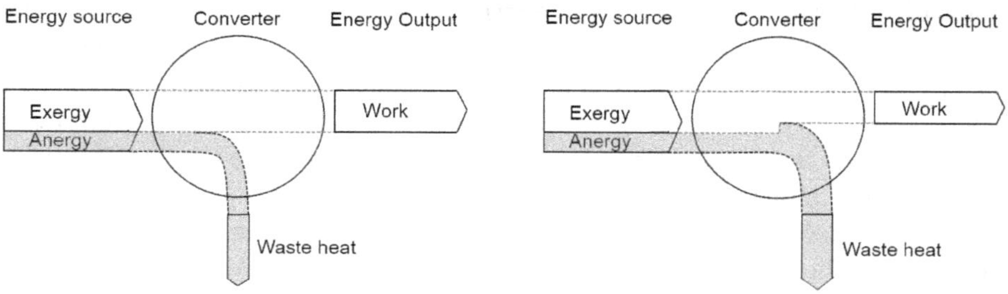

Fig. 1.5: Energy and exergy flow through an ideal converter (*left*) and a realistic converter where irreversibilities reduce the amount of work that can be performed (*right*). The width of the arrows is proportional to the amount of energy in the respective forms.

The example on the left represents an ideal, reversible converter, such as a Carnot machine, that transforms the exergy part of the input energy fully into the desired work while the anergy part is rejected to its surroundings as waste heat. In contrast to this, the right-hand example depicts the case of a realistic converter where irreversible losses occur within the device converting a part of the input exergy into waste heat. The first law efficiency of the conversion processes are the ratios between the widths of the arrow denoted "work" on the right-hand side of both diagrams and the total widths of the arrow on the left-hand side of the diagrams. Even for the ideal process on the left $\eta_1 < 1$ holds. In the realistic case, the first law efficiency is the product of the exergy content of the total energy entering the converter and the quality of the conversion process itself. It is thus a convolution of two effects, the ratio between the exergy content Φ_{in} and the total input energy E_{in} on the one hand, and the quality of the conversion process on the other hand. For an ideal process, which is by definition fully reversible, the first law efficiency is a useful measure in itself. It allows for an evaluation of the feasibility of the conversion process chosen. Any ideal reversible process, however, runs infinitely slowly as both directions of the process are always proceeding at an identical speed. This means that every real device needs a driving force which forces the process to run in one direction at a finite net speed and introduces an irreversibility. An efficient process runs at the desired speed with low losses due to the necessary irreversibilities. In turn, the irreversibilities are directly linked to exergy losses and convert exergy to anergy. The relevant quantity to assess the quality of the engineering of a given energy conversion process is thus the amount of exergy that is preserved during the conversion process. This quantity is called the **exergy or second-law efficiency**, η_2, as it accounts for the second law of thermodynamics as given below in equation (2.20):

$$\eta_2 = \frac{\Phi_{\text{out}}}{\Phi_{\text{in}}}, \tag{1.6}$$

where $\Phi_{\text{in,out}}$ are the input and output exergies, respectively. For the assessment of the energy conversion processes in this book, the dependence of the second-law efficiency on the driving force will be the method of choice. The development of a deep physical understanding of the actual driving mechanism will thus take a central role.

1.2.3 Example: heating of a room

A good example elucidating the difference between first- and second-law efficiency is the heating of a room with electrical energy. Electrical energy is pure exergy, as will be shown below in Chapter 5. The heating can be done in two different ways: first, by directly converting the electrical energy into heat, and, second, by using the electrical energy to drive an electrical heat pump which generates a heat flow from a lower temperature level to a higher temperature level. The exergy flow diagrams for both heating devices are shown in Figure 1.6.

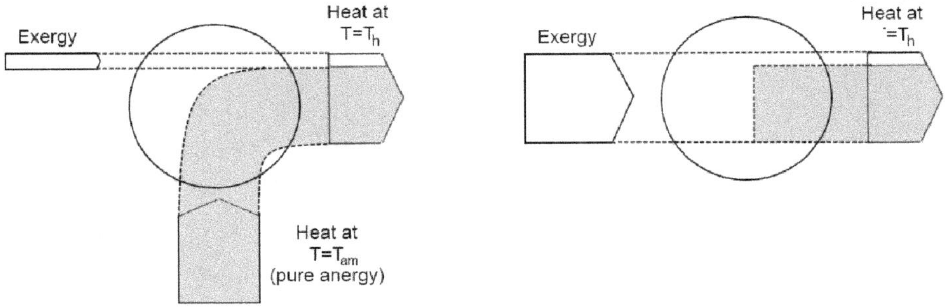

Fig. 1.6: Energy and exergy flow diagram for electrical heating with a heat pump (*left*) and direct electrical heating (*right*). The heat rejection (*right*) and heating temperatures of both devices are identical, leading to an identical energy and exergy output.

During the direct electrical heating no energy is converted into an undesired form as all the electrical input energy is converted to heat leading to a first law efficiency of

$$\eta_1 = \frac{Q_{\text{out}}}{E_{\text{el,in}}} = 1. \tag{1.7}$$

Given this value, one might come to the conclusion that this form of heating is indeed already ideal. However, while the energy is conserved, it is converted from a form of energy with a high exergy content, i.e. electrical energy, to a form of energy with low exergy content, namely heat provided at $T = T_{\text{h}} \approx 60\,^{\circ}\text{C}$. The second-law efficiency is consequently much lower:

$$\eta_2 = \frac{\Phi_{\text{out}}}{\Phi_{\text{in}}} = \frac{\Phi_{\text{out}}}{E_{\text{el,in}}} \ll 1 \;\; (\approx 10\,\% \text{ at } T_{\text{h}} = 60\,^{\circ}\text{C}). \tag{1.8}$$

Conversely, the heat pump does not convert the input exergy partly into anergy but makes use of it to extract heat from the ambient surroundings. The input exergy is thereby "diluted" in a way that the pumped anergy together with the input exergy is exactly constituting the exergy and anergy content of the heat rejected to the heated room. Since all processes are assumed to be reversible, no exergy is destroyed, leading to

$$\eta_2 = \frac{\Phi_{\text{out}}}{\Phi_{\text{in}}} = 1. \tag{1.9}$$

The first law efficiency is larger than one in this case, as the anergy taken from the environment, which is an infinite heat bath, is not counted as input energy:

$$\eta_1 = \frac{Q_{\text{out}}}{E_{\text{el,in}}} > 1. \tag{1.10}$$

This example shows the usefulness of the second-law efficiency for the evaluation of energy conversion processes. A good energy conversion process is characterized by preserving a large part of the input exergy in the exergy content of the output energy.

2 Exergy in closed systems

Energy considerations have always to be seen within the framework of thermodynamics. Only in this context can energy conversion processes in macroscopic systems be appropriately described. Accordingly, in the beginning of this chapter a short recapitulation of basic terms and concepts of thermodynamics are discussed. More details can be found in Appendix A.

2.1 Thermodynamic basics

In thermodynamics, the subject of our study, e.g. a gas, a fuel cell, or an engine, is called the **thermodynamic system.** It is separated from its surroundings by clearly defined borders and can interact with its surroundings through these borders. Depending on the interactions with the surroundings, three types of thermodynamic systems are distinguished. An **isolated system** does not interact with its surroundings at all, a **closed system** can exchange heat and work with its surroundings, and an **open system** can also exchange matter. The three types of thermodynamic systems are schematically shown in Figure 2.1.

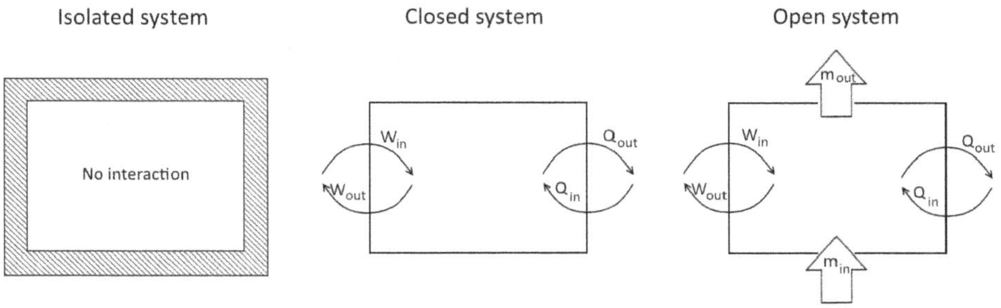

Fig. 2.1: Sketch of the three different types of thermodynamical systems: an isolated system (*left*) does not interact with its surroundings; a closed system (*center*) can exchange heat and work with its surroundings; and an open system (*right*) can exchange heat, work, and matter with its surroundings.

The state of the system is characterized by state variables, such as temperature, T, pressure, p, or volume, V. It is convenient to distinguish between extensive and intensive state variables where extensive state variables, as for example the volume, are proportional to the system size, while intensive state variables are independent of the system size as, e.g. the temperature. The change of the state of a system is called a **process.** Typically considered model processes for state changes are **isothermal** (T stays constant), **isobaric** (p stays constant), **isentropic** (S stays constant), **iso-**

choric (V stays constant), or **adiabatic** (no exchange of heat across the borders of the system). For all ideal processes, isentropic and adiabatic state changes are identical. A detailed description of these state changes in closed systems can be found in Section A.2.2 of the appendix.

The first important distinction of different types of energy is the distinction between energy as a state variable and energy which is transferred across the borders of a system during a state change. The former types of energy are for example the thermodynamic potentials **internal energy**, U, **enthalpy**, H, **Helmholtz free energy**, F, and **Gibbs free energy**, G. In this book, such energies are written with an index denoting the state of the system, e.g. U_1 for the internal energy of state '1'. For these quantities it is sensible to use expressions as, e.g. $\Delta U = U_{end} - U_{in}$, as they take on one well-defined value for each state of a given thermodynamic system. In contrast, energy transferred across the borders of a system exists only during the process itself. Thus, it does not characterize the state of a system and can therefore not be expressed as a difference between two values. Rather, the convention chosen in this book is to label it with left and right subindices for the initial and final state of the system. An amount of energy transferred across the border of a system while it changes from state '1' to state '2', is thus denoted by ${}_1E_2$. Energy only crosses the borders of a system when the system is not in equilibrium with its surrounding and a driving force for the energy across the border of the system exists. The energy transferred across the borders of a closed system is then characterized by the origin of this driving force and can take the two forms **heat**, Q, and **work**, W. Heat is energy that crosses the borders of a system due to a temperature gradient across these borders and work performed on or by the system is caused by various possible driving forces including pressure gradients, gravitational force or electric fields.

2.1.1 The first law of thermodynamics

The first law of thermodynamics is concerned with changes of the energy of a system. For a closed system, it can be formulated as follows:

! **The internal energy of a (closed) system only changes if heat or work are transferred across the borders of the system:**

$$\begin{aligned} \Delta U &= Q + W \\ dU &= \delta Q + \delta W \end{aligned} \tag{2.1}$$

This law is a form of the energy conservation principle, as it states that energy can neither be destroyed nor created but merely transported between thermodynamic systems and converted between different forms. The quantity U is the internal energy of the closed system, i.e. the sum of all kinetic and potential energies of all particles in the system. It is an extensive state variable. The respective shares of heat and work in the total energy transfer upon a state change are path dependent. Note again that while the term ΔU makes sense as the system changes from one value of the internal energy to another due to the energy transfer across its border, a term of the form ΔQ or ΔW would not be a sensible choice, as neither quantities are state variables, and it is hence impossible to assign an initial or final amount of heat or work to the system. Terms of that form do, however, unfortunately often appear in textbooks and give rise to the incorrect impression there could be a meaningful concept of "heat or work contained in a system", which is not the case. Heat and work are no state variables, and such a notion is hence meaningless.

If the first law of thermodynamics is formulated as in equation (2.1), this implies the following sign convention for heat and work, which will be kept throughout this book:

Heat and work have a positive sign if they are added to the system and a negative sign if they are provided by the system.

In differential form for reversible processes the first law of thermodynamics reads

$$dU = \underbrace{TdS}_{\delta Q} - \underbrace{pdV}_{-\delta W}, \tag{2.2}$$

where the δ indicates that the differentials δQ and δW are not exact. The sign of $\delta W = -pdV$ is due to the sign convention mentioned. If the system expands, i.e. $dV > 0$, it performs work at its surroundings which leads to a negative work performed at the system.

2.1.2 The second law of thermodynamics

The second law of thermodynamics is a statement about the possibility or impossibility of the spontaneous occurrence of certain processes. It exists in several different formulations, which are all equivalent and each of which is useful in a different context. While the first formulation given below is most useful when considering basic physical principles, the fourth formulation is most useful for the study of heat engines. For a more qualitative feeling of the implications of the second law of thermodynamics, the other two formulations are typically helpful.

The first formulation reads:

The entropy of an isolated system cannot decrease in any state change. It does not change in reversible processes.

$$\Delta S \geq 0; \quad \Delta S = 0 \Leftrightarrow \text{reversible process.} \tag{2.3}$$

Note that in a closed system, as opposed to an isolated system, the entropy can decrease when the system rejects heat reversibly to the surroundings.

The second formulation was given by Clausius:

There is no thermodynamic state change where the only consequence is the transfer of heat from a heat bath at lower temperature to a heat bath at higher temperature.

This formulation captures the everyday experience that heat only flows spontaneously from a body at higher temperature to a body at lower temperature. Such a process is irreversible, and the entropy of the system encapsulating both heat baths increases in all processes allowed under the second law of thermodynamics in this formulation. The overall system is then not exchanging heat with its surroundings, making it an isolated system, which leads back to the first formulation of the second law of thermodynamics.

The third formulation was given by Kelvin and Planck:

It is impossible to construct a perpetuum mobile of the second kind, i.e. a periodically working machine that has no other effects than to lift a weight and cool a heat bath.

This means that heat provided to a heat engine cannot fully be transformed into work of this engine, which would indeed also lead to a violation of the two formulations of the second law of thermodynamics given above.

It is possible to find an optimal cycle to transform a heat flow from a higher to a lower temperature level into work by a heat engine. The corresponding thermodynamic cycle is called the **Carnot cycle**, which leads to the fourth equivalent formulation of the second law of thermodynamics:

The efficiency of a heat engine exclusively driven by a heat flow from a heat bath at temperature T_h to a heat bath at temperature $T_c < T_h$ cannot exceed the Carnot efficiency of

$$\eta_C = \left(1 - \frac{T_c}{T_h}\right) \tag{2.4}$$

In the course of this chapter a fifth formulation will be given.

2.2 Exergy in closed systems

The definition of exergy given in the last chapter requires some elaboration. For this reason a more rigorous definition and discussion of exergy in closed systems is given in this section. Exergy is defined in the following way:

The exergy Φ_i of a system in state 'i' is the maximal possible amount of useful work the system can perform when it is brought into equilibrium with the ambient state.

To calculate the exergy of a given state one has to construct a path from the initial state to the ambient state that maximizes the useful work the system performs during the state change. In doing so, care has to be taken that no further exergy is added to the system while it moves along the chosen path. Heat transferred at ambient temperature is pure anergy, and heat transferred at all temperatures different from the ambient temperature carries exergy (for details see Section 2.3.2). This means that on the path chosen any heat transfer has to occur isothermally at ambient temperature. As isothermal processes are not changing the temperature of the system, processes that do not involve heat transfers, i.e. adiabatic processes, have to be used to adapt the temperature of the system. Furthermore, the theoretically maximal work is obtained in a reversible process. Hence the adiabatic process is also an isentropic process. These considerations leave only one possible route from the initial state to the ambient state. It consist of an initial adiabatic expansion whereby the temperature is decreased from the initial temperature $T = T_i$ to ambient temperature $T = T_{am}$, with the resulting intermediate pressure p_{im} still being larger than p_{am}, and the volume V_{im} thus still being smaller than when the system is in equilibrium with its surroundings. This is followed by an isothermal expansion at $T = T_{am}$ until ambient pressure is reached. The processes described are depicted in Figure 2.2.

In the following, the useful work the system can perform during these two processes is considered separately for the two processes. During the isentropic expansion from the initial state 'i' to the intermediate state 'im' the heat transferred is $\delta Q = 0$,

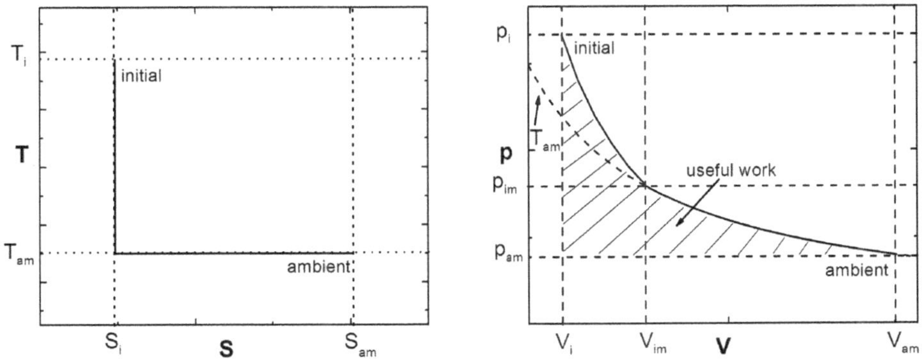

Fig. 2.2: TS- and pV-diagrams of the fully reversible state transitions from the initial state to the ambient state consisting of an initial isentropic expansion to adjust the temperature to $T = T_{am}$, followed by an isothermal expansion at $T = T_{am}$. The useful work extracted in the process is marked in the pV-diagram.

and, according to the first law of thermodynamics the following expression holds:

$$U_{\mathrm{im}} - U_{\mathrm{i}} = -\int_{V_{\mathrm{i}}}^{V_{\mathrm{im}}} p dV = {}_{\mathrm{i}}W_{\mathrm{im}}. \tag{2.5}$$

A part of this work, the **work of displacement** W^{dis}, is required to expand the system against the ambient pressure of the surroundings, and only the remaining part is exchanged in the form of **useful work**, i.e. work that can be used for lifting a weight or rotating a shaft. Thus, the following expression holds for the useful work exchanged during the process:

$$_{\mathrm{i}}W_{\mathrm{im}} =: {}_{\mathrm{i}}W_{\mathrm{im}}^{\mathrm{use}} + {}_{\mathrm{i}}W_{\mathrm{im}}^{\mathrm{dis}} = {}_{\mathrm{i}}W_{\mathrm{im}}^{\mathrm{use}} - p_{\mathrm{am}}(V_{\mathrm{im}} - V_{\mathrm{i}}). \tag{2.6}$$

Combining equations (2.5) and (2.6) the following expression is obtained:

$$_{\mathrm{i}}W_{\mathrm{im}}^{\mathrm{use}} = U_{\mathrm{im}} - U_{\mathrm{i}} + p_{\mathrm{am}}(V_{\mathrm{im}} - V_{\mathrm{i}}). \tag{2.7}$$

In the intermediate state 'im' the system is still at an elevated pressure and can thus perform mechanical work against the ambient state. After reaching thermal equilibrium with the surroundings in the first step just discussed, in the second step it is brought into mechanical equilibrium with the surroundings in an isothermal process at $T = T_{\mathrm{am}}$. The heat $_{\mathrm{im}}Q_{\mathrm{am}}$ reversibly transferred in this process is given by

$$_{\mathrm{im}}Q_{\mathrm{am}} = T_{\mathrm{am}}(S_{\mathrm{am}} - S_{\mathrm{im}}) = T_{\mathrm{am}}(S_{\mathrm{am}} - S_{\mathrm{i}}), \tag{2.8}$$

where the second equality holds as the states 'i' and 'im' are connected by an isentropic state change. According to equation (2.8) and the first law of thermodynamics, equation (2.1), the useful work exchanged in this process is then given by

$$_{\mathrm{im}}W_{\mathrm{am}}^{\mathrm{use}} = U_{\mathrm{am}} - U_{\mathrm{im}} + p_{\mathrm{am}}(V_{\mathrm{am}} - V_{\mathrm{im}}) - T_{\mathrm{am}}(S_{\mathrm{am}} - S_{\mathrm{i}}). \tag{2.9}$$

For the useful work exchanged during the succession of the two processes, i.e. when the system is brought from the initial state 'i' into thermal and mechanical equilibrium with the ambient surroundings, the following expression is thus obtained:

$$\begin{aligned} _{\mathrm{i}}W_{\mathrm{am}}^{\mathrm{use}} &= {}_{\mathrm{i}}W_{\mathrm{im}}^{\mathrm{use}} + {}_{\mathrm{im}}W_{\mathrm{am}}^{\mathrm{use}} \\ &= U_{\mathrm{im}} - U_{\mathrm{i}} + p_{\mathrm{am}}(V_{\mathrm{im}} - V_{\mathrm{i}}) + \\ &\quad + U_{\mathrm{am}} - U_{\mathrm{im}} + p_{\mathrm{am}}(V_{\mathrm{am}} - V_{\mathrm{im}}) - T_{\mathrm{am}}(S_{\mathrm{am}} - S_{\mathrm{i}}) \\ &= (U_{\mathrm{am}} - U_{\mathrm{i}}) + p_{\mathrm{am}}(V_{\mathrm{am}} - V_{\mathrm{i}}) - T_{\mathrm{am}}(S_{\mathrm{am}} - S_{\mathrm{i}}). \end{aligned} \tag{2.10}$$

The exergy of state 'i' should be positive when the system performs work during the state changes just considered, while the sign of the useful work is negative due to the sign convention used. This leads to the following expression for the exergy of a closed system in state 'i':

$$\Phi_{\mathrm{i}} = -{}_{\mathrm{i}}W_{\mathrm{am}}^{\mathrm{use}} = (U_{\mathrm{i}} - U_{\mathrm{am}}) - p_{\mathrm{am}}(V_{\mathrm{am}} - V_{\mathrm{i}}) + T_{\mathrm{am}}(S_{\mathrm{am}} - S_{\mathrm{i}}). \tag{2.11}$$

Thus, the exergy of state 'i' can be calculated by considering state variables only. The difference in internal energies between state 'i' and the ambient state has to be corrected by two terms. First, the work against the ambient pressure necessary to displace the surroundings while expanding the system from its initial to its final volume has to be subtracted, and, second, the pure anergy (heat) which is transferred into the system by the ambient surroundings during the isothermal part of the expansion to adapt the entropies of both states has to be added. So far, a system at rest and on the height level of the surroundings has been implicitly assumed. Whenever this is not the case, kinetic energy and potential energy of the system have to be added to the internal energy of the initial state. Then these two forms of mechanical energy are counted as pure exergy, a notion that will be justified below.

To simplify calculations it is now possible to rearrange the terms of equation (2.11) in such a way that the extensive variables associated with the ambient state are separated. With this the **exergy function** of the initial state, Φ_i^x, can be defined in the following way:

$$\Phi_i = \underbrace{(U_i + p_{am} V_i - T_{am} S_i)}_{=:\Phi_i^x} - (U_{am} + p_{am} V_{am} - T_{am} S_{am}), \tag{2.12}$$

and the exergy of state 'i' can be restated in terms of exergy functions:

$$\Phi_i = \Phi_i^x - \Phi_{am}^x. \tag{2.13}$$

It is easy to see from equation (2.13) that for any two states '1' and '2' the differences in exergy and exergy function are identical:

$$\Phi_2 - \Phi_1 = \Phi_2^x - \Phi_1^x. \tag{2.14}$$

In most cases when differences are concerned it is easier to calculate the differences in exergy functions rather than the differences in exergies.

2.3 Exergy transfers

When energy is transferred across a border of a system, the system undergoes a state change which also changes the exergy of the system. If the energy transfer process is reversible, the exergy loss of the source system is identical to the exergy gain of the target system. Conversely, for irreversible energy transfer processes the former exceeds the latter, and the second-law efficiency of the energy transfer process, i.e. the ratio of both, is smaller than 1. In this book the energy is extracted from its sources and given to the converters/consumers in one of the following four forms: **mechanical work**, **heat transfers** at a given temperature, **electricity**, and energy stored in **chemical bonds**. The exergy content of the former two types of energy will be discussed in the following sections while large parts of Chaps. 5 and 6 will be devoted to the exergy content of the latter two.

2.3.1 Exergy transfer via work transfer

In isentropic processes only work is transferred across the border of a given system. In this case, the exergy change of the system and consequently the exergy content of the transferred work $\Phi({}_1W_2)$ is given by:

$$\begin{aligned}\Phi({}_1W_2) = \Phi_2 - \Phi_1 &= \Phi_2^{\mathrm{x}} - \Phi_1^{\mathrm{x}} = (U_2 - U_1) - \underbrace{T_{\mathrm{am}}(S_2 - S_1)}_{=0} + p_{\mathrm{am}}(V_2 - V_1) \\ &= {}_1W_2 + p_{\mathrm{am}}(V_2 - V_1) = -\int_{V_1}^{V_2} p dV + p_{\mathrm{am}}(V_2 - V_1) = {}_1W_2^{\mathrm{use}}.\end{aligned} \tag{2.15}$$

The process is visualized in Figure 2.3.

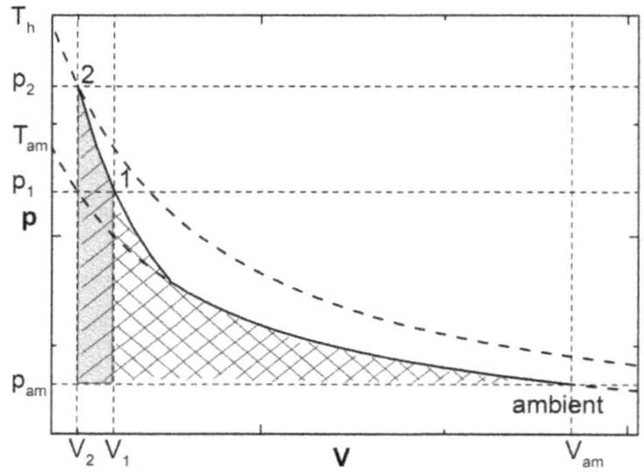

Fig. 2.3: pV-diagram of a system undergoing a state change 1 → 2 by an isentropic work performance. The respective exergies of the states 1 (*slanted-left lines*), 2 (*slanted-right lines*) and the exergy content of the work taken up by the system (*gray*) are the marked areas.

This means that for a closed system the amounts of energy and exergy transferred in form of work across the borders of the system during the state change of the system differ by the amount of work required to displace the surroundings. The useful part of the work ${}_1W_2^{\mathrm{use}}$ transferred across the borders of the system is hence the exergy content of the work transferred.

2.3.2 Exergy transfer via heat transfer at $T = T_{\mathrm{h}}$

To determine the exergy content of heat, the exergy change of a system taking up the heat ${}_1Q_2$ at $T = T_{\mathrm{h}}$ in an isothermal transfer of heat has to be determined. Note at the outset that it is in general not possible to just transfer heat isothermally without a corresponding work performance at the system. As the system receives heat isothermally, it performs work at its surroundings, its useful part being $-{}_1W_2^{\mathrm{use}} = -{}_1W_2 - p_{\mathrm{am}}(V_2 - V_1)$. The exergy transferred by the heat itself is therefore by an amount $-{}_1W_2^{\mathrm{use}}$ larger than the difference in the exergy functions $\Phi_2^{\mathrm{x}} - \Phi_1^{\mathrm{x}}$. Using equation (2.14) and the first law of thermodynamics, equation (2.1), the exergy $\Phi({}_1Q_2)$ can then be calculated in the following way:

$$\begin{aligned}\Phi({}_1Q_2) &= \Phi_2^{\mathrm{x}} - \Phi_1^{\mathrm{x}} - {}_1W_2^{\mathrm{use}} = \Delta U + p_{\mathrm{am}}\Delta V - T_{\mathrm{am}}\Delta S - {}_1W_2^{\mathrm{use}} \\ &= {}_1W_2 + {}_1Q_2 - T_{\mathrm{am}}\Delta S + \underbrace{p_{\mathrm{am}}\Delta V - {}_1W_2^{\mathrm{use}}}_{=-{}_1W_2} \\ &= {}_1Q_2 - T_{\mathrm{am}}\Delta S = \left(1 - \frac{T_{\mathrm{am}}\Delta S}{{}_1Q_2}\right)\cdot {}_1Q_2 = \left(1 - \frac{T_{\mathrm{am}}}{T_{\mathrm{h}}}\right)\cdot {}_1Q_2\end{aligned} \tag{2.16}$$

For an isothermal heat transfer at $T = T_{\mathrm{h}}$ the exergy thus corresponds to the transferred heat multiplied with the Carnot efficiency of a heat engine working between the transfer temperature T_{h} and the ambient temperature T_{am}.

As an example, the processes described are visualized for an ideal gas as the working medium in Figure 2.4. In this case, the internal energy stays constant in isothermal processes, i.e. $\Delta U = 0$. During the heat transfer the system is thus expanding while performing the work ${}_1W_2 = -{}_1Q_2$. For the exergy $\Phi({}_1Q_2)$ transferred into the system via the heat transfer this leads to the following expression.

$$\Phi({}_1Q_2) + \Phi({}_1W_2^{\mathrm{use}}) = \Phi_2 - \Phi_1 \quad \Rightarrow \quad \Phi({}_1Q_2) = (\Phi_2 - \Phi_1) - {}_1W_2^{\mathrm{use}}. \tag{2.17}$$

Note that $\Phi({}_1W_2^{\mathrm{use}})$ is here negative, as it represents exergy leaving the system.

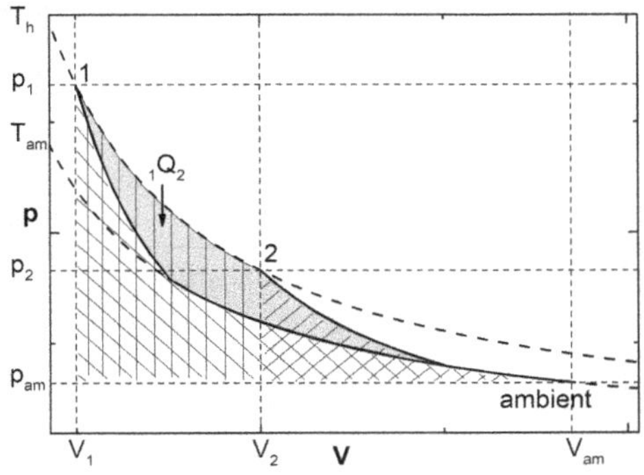

Fig. 2.4: pV-diagram of a system undergoing a state change 1 → 2 by an isothermal heat injection at $T = T_{\mathrm{h}}$. The respective exergies of the states 1 (*slanted-left lined area*) and 2 (*slanted-right lined area*) are the marked areas. During the process useful work is performed by the system (*vertical lined area*) which has to be subtracted from the exergy difference to find the exergy content of the heat taken up by the system (*gray area*).

In Figure 2.4 the exergies of state 1 (*slanted-left lined area*) and 2 (*slanted right lined area*) are depicted together with the useful work released by the system during the process (*vertically lined area*). According to equation (2.17) the exergy $\Phi({}_1Q_2)$ carried by the heat into the system is marked in Figure 2.4 by *vertical + slanted-right – slanted-left lined area*. The thus marked area, highlighted in *gray*, then indeed corresponds to a Carnot cycle between the temperature levels T_{h} and T_{am}. A detailed description of the Carnot cycle can be found in Section A.3 of the appendix.

2.3.3 Exergy and anergy contents of transferred energy

The exergy content of useful work is 100 %, and the exergy content of heat transferred at a given temperature is exactly the amount of useful work that can be gained when the heat is driving a Carnot machine, i.e. an optimal converter from heat to useful work. From this a definition for the exergy content of the transferred energy can be derived:

The exergy content of energy transferred across the border of a system is the maximal amount of useful work that can be generated by this energy in an optimal process.

Recapitulating equation (2.11), the anergy content $A(E)$ of energy $E = Q + W$ fed to a system causing a state change from state '1' to state '2' can be defined as

$$\begin{aligned} A(E) &= E - \Phi(E) = E - (\Phi_2 - \Phi_1) = E - (\Phi_2^{\mathrm{x}} - \Phi_1^{\mathrm{x}}) \\ &= E - \underbrace{\Delta U}_{=E} - p_{\mathrm{am}}\Delta V + T_{\mathrm{am}}\Delta S = -p_{\mathrm{am}}\Delta V + T_{\mathrm{am}}\Delta S, \end{aligned} \tag{2.18}$$

where the identity $E = \Delta U$ holds due to the first law of thermodynamics, equation (2.1). The anergy content of energy crossing the border of a system is hence the minimal amount of energy required to restore the original system volume and entropy. This minimal energy required for restoring the initial state has to be factored in for a correct assessment of the total useful energy that can be gained in an optimal process. If it were neglected, the optimal converter would be allowed to continuously increase its entropy or change its size, and the maximal amount of useful work would be overestimated. Linking equation (2.18) to the definition of the exergy content of energy fed to a system, the anergy content of energy transferred to a system can hence be stated as follows:

The anergy content of energy crossing the border of a system is the sum of the minimal amounts of energy required to discard the entropy increase and revert the volume change of the system upon the state change accompanying the energy transfer.

The concept of an exergy and anergy content of energy transferred across the borders of a system is useful, as it is not linked to a specific thermodynamic system and allows for the comparison of different conversion routes of the energy in question.

2.3.4 The laws of thermodynamics in terms of exergy

Exergy and anergy can be both seen as state variables, as they only depend on the present state of a given system and the fixed ambient state. It is thus often useful to reformulate the energy exchange between thermodynamic systems in terms of exergy and anergy exchanges rather than heat and work transfers. The first law of thermodynamics can then be reformulated in the following way:

$$\Delta U = \Delta\Phi + \Delta A = \Phi(E) + A(E), \tag{2.19}$$

where the first equality considers state variables only, while the second equality links the changes in exergy and anergy within the system to the exergy and anergy contents of the energy transferred across the borders of the system during the state change. In this formulation both Δ's at the right-hand side of the first equality are sensible terms, as the initial and the final states of the closed system contain a defined amount of exergy and anergy, respectively. Most importantly, the respective shares of exergy and anergy of the energy transferred are identical to $\Delta\Phi$ and ΔA, respectively, and determine the initial and the final state of the system and are thus path independent.

Exergy is destroyed in any irreversible process and converted to anergy as the entropy increases in any irreversible process. At the same time, exergy cannot be produced within an isolated system, which leads to a fifth equivalent formulation of the second law of thermodynamics:

The exergy content of an isolated system cannot increase. It is decreased by any irreversible process: !

$$\Delta\Phi \leq 0; \quad \Delta\Phi = 0 \Leftrightarrow \text{reversible process} \tag{2.20}$$

In the field of energy science, the formulations of the first and second law of thermodynamics given by equations (2.19) and (2.20) are very useful, as will be demonstrated throughout the rest of this book.

2.4 Exergy sources

On the global scale, the main source of exergy is the solar irradiation, which is mostly exergy, as will be discussed in detail in Section 8.1.2. All sorts of energy conversion processes happening within the atmosphere of the earth lead to a depletion of this exergy input. These processes include the water cycle, movement of air, and biological processes, all of which in turn provide renewable exergy sources. The earth then emits essentially pure anergy in the form of low temperature heat radiation into space. Neglecting long term fluctuations, the total net influx of energy is zero. This has to be the case, as the earth would otherwise either be cooling or warming.

Apart from the solar irradiation, two additional, albeit several orders of magnitude smaller, sources of renewable exergy exist: first, geothermal exergy stemming from the hot inside of the earth, and, second, gravitational energy from both moon and sun. While the geothermal energy reaching the surface of the earth is pure anergy, at the surface of the earth it has a significant exergy content in deeper layers of the crust of the earth, as the temperature is higher there. The source of this heat is in roughly equal parts the friction heat generated by the colliding objects which formed our planet over 4 billion years ago and radioactive decays inside the earth.

The gravitational fields of moon and sun lead to tidal waves in the oceans, the amplitude of which constitutes exergy in the form of potential energy in the gravitational field of the earth. The exergy the surface of the earth receives from the sun as well as from its core and the gravitational fields of moon and sun are stable over timescales that can be treated as infinite in any meaningful context for human technology. These three sources of exergy are thus fully renewable.

Before their radioactive decay, the unstable isotopes in the crust of the earth offer an additional source of exergy that can be mined and used in nuclear power converters, albeit with considerable effort associated with the mining and refining of the isotopes.

Lastly, another source of exergy are the organic compounds stored within the sedimentary layers of the crust of the earth due to photosynthetic processes in the course of hundreds of millions of years. Although these organic compounds are both finite and an almost negligible part of the total amount of exergy available to humanity, they currently account for most of our exergy consumption (81 % of the primary energy consumption in 2010).

2.5 Heat pumps

Heat pumps provide good examples for exergy considerations in real devices and show the value of exergy flow diagrams. A heat pump is a device that is driven by energy with a high exergy content and transfers heat from a low temperature level $T = T_{am}$ to a high temperature level $T = T_h$. By extracting energy from a low temperature level and rejecting it as heat at a higher temperature level it also cools the cold heat bath, and consequently the same principle is used in most cooling devices such as refrigerators and air conditioning units. There are two main types of heat pumps: compression heat pumps driven by electrical energy, and absorption heat pumps driven by heat provided at a relatively high temperature level $T = T_d > T_h$. The exergy flow diagrams for the two different types of heat pumps are shown in Figure 2.5.

It is easy to see that the energy output exceeds the energy input for both types of heat pumps shown in Figure 2.5, leading to a first-law efficiency greater than one. The energy output has a lower exergy content than the input. The total energy is thus increased by "diluting" the exergy input with pure anergy taken from the ambient surroundings. One could also say that the anergy taken from the ambient surroundings is upgraded to a useful type of energy by the exergy driving the heat pump. The second-law efficiency η_2 is smaller than 1 for both processes shown in Figure 2.5, as conversion

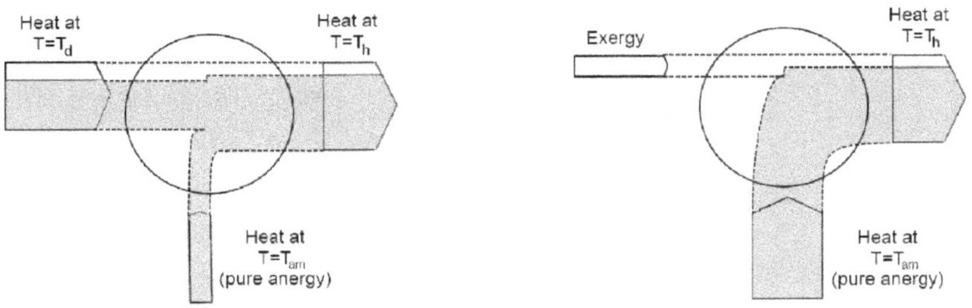

Fig. 2.5: Energy and exergy flow diagrams of an absorption heat pump (*left*) and a compression heat pump (*right*) with some irreversibilities of the pumps included. Both pumps have the same output: heat transferred at $T = T_h$.

losses in the heat pump are already accounted for. The second-law efficiency is only changed by these conversion losses and is hence a measure of the quality of the heat pumps. With the exergy content of heat transferred at a given temperature T_h, as calculated in equation (2.16), the following value for the first-law efficiency is obtained:

$$\eta_1 = \frac{Q_{\text{out}}}{E_{\text{in}}} = \frac{\Phi_{\text{out}}}{\Phi_{\text{in}}} \cdot \frac{\Phi_{\text{in}}}{E_{\text{in}}} \cdot \frac{Q_{\text{out}}}{\Phi_{\text{out}}} = \eta_2 \cdot \frac{\Phi_{\text{in}}}{E_{\text{in}}} \cdot \left(\frac{1}{1 - T_{\text{am}}/T_{\text{h}}} \right), \tag{2.21}$$

where $\Phi_{\text{in}}/E_{\text{in}}$ is the exergy content of the input energy. For the electrical heat pump the input energy is pure exergy, and thus the following expression holds:

$$\eta_1 = \eta_2 \cdot \left(\frac{1}{1 - T_{\text{am}}/T_{\text{h}}} \right) > 1. \tag{2.22}$$

In the case of the absorption heat pump, $\Phi_{\text{in}}/E_{\text{in}} = (1 - T_{\text{am}}/T_{\text{D}})$ holds, leading to

$$\eta_1 = \eta_2 \cdot \left(\frac{1 - T_{\text{am}}/T_{\text{d}}}{1 - T_{\text{am}}/T_{\text{h}}} \right) > 1. \tag{2.23}$$

The lower the temperature of the output energy, the lower its exergy content becomes. This means that more anergy can be added to the input energy, increasing the first-law efficiency of the heat pump.

2.5.1 Working principle

In Figure 2.6 a schematic of a heat pump is shown. The working fluid of a heat pump undergoes a phase transition during both heat transfer processes. Both heat exchanges are taking place at different pressure levels, and the key to the working principle of a heat pump is the pressure dependence of the boiling temperature $T_{\text{boil}}(p)$ of the working fluid. With this pressure dependence it is possible to condense the working fluid at a high pressure level at a temperature $T = T_h$ higher than the boiling temperature $T = T_c$ at a lower pressure level. Evaporation heat is thereby taken into

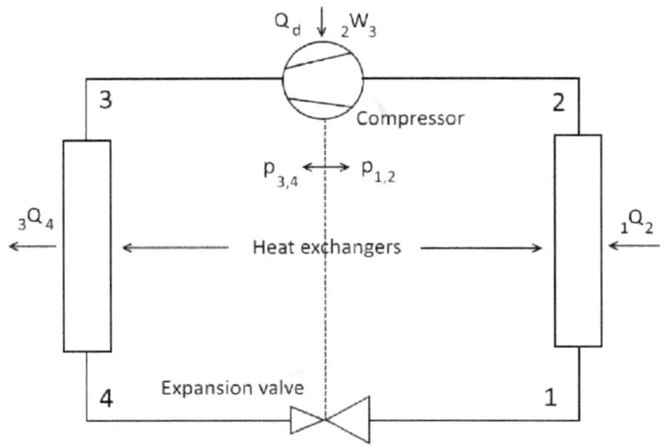

Fig. 2.6: Schematic of a heat pump. In an electrical heat pump the compressor is electrically driven, and in an absorption heat pump it works with the driving heat at $T = T_d$. All other parts are essentially identical for both types of heat pumps. The *vertical dashed line* separates the two pressure levels in the cycle.

the working fluid at a low temperature level $T = T_{am}$ and released at a high temperature level $T = T_h$. Details concerning phase transitions are given in Section A.4 of the appendix. The TS- and pV-diagrams of a heat pump corresponding to the schematic shown in Figure 2.6, and the corresponding numeration of the thermodynamic states are shown in Figure 2.7.

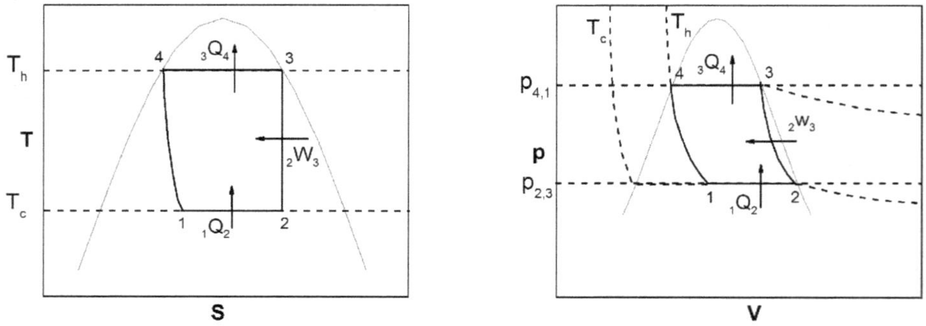

Fig. 2.7: TS- and pV-diagrams of the heat pump cycle. The area of coexisting phases (below the *gray line*) is shown together with the isothermal, adiabatic, and isenthalpic processes involved.

1 → 2: At the lower, ambient temperature $T = T_{am}$ the working fluid is kept at a pressure where the boiling point is lower than ambient temperature and consequently it undergoes a phase change from the liquid to the gaseous phase taking up the evaporation heat $_1Q_2$ from the surrounding heat bath at $T = T_{am}$. The process takes place in the region where both phases coexist and is therefore both isobaric and isothermal (see Section A.4), and only the steam mass fraction x of the working fluid increases.

2 → 3: In this step the gas is actively compressed, utilizing the work ${}_2W_3$ and increasing the pressure to the higher pressure level $p = p_{3,4}$ and the boiling point to $T = T_\mathrm{h}$. In absorption heat pumps, the driving heat at $T = T_\mathrm{d}$ is used for the compression of the working fluid.

3 → 4: At the higher pressure level the gas condenses at an increased boiling point at $T = T_\mathrm{h}$ releasing the condensation heat ${}_3Q_4 < 0$ to the surroundings. As this process again takes place within the region of coexisting phases, the process is both isothermal and isobaric, and only the steam mass fraction x decreases.

4 → 1: In the last step, the working fluid is relaxed to the initial pressure, which corresponds to an isenthalpic process in an expansion valve. This process is irreversible.

Apart from the last irreversible step 4 → 1 the heat pump realizes a Carnot cycle.

3 Exergy in open systems

In many realizations of thermodynamic cycles, a working fluid flows through different components of an engine in which it performs the individual steps of the cycle. To analyze the individual components of such a thermodynamic cycle, open systems, as opposed to closed systems, have to be considered. In a closed system the working fluid is a constant amount of matter confined to a volume with moveable borders. In contrast, in an open system a mass current flows through a fixed volume, the so-called control volume. In addition to heat and work, matter is also transferred across the borders of the system. A closed system cannot be operated in a continuous manner, as the volume has to change periodically in time. In an open system, in contrast, the state at any position in the control volume can become stationary in time, and only the working fluid undergoes state changes while passing through the control volume. This makes a continuous operation possible. In this chapter, first a short overview of thermodynamics in flow systems will be given, and then a special case, the steady flow equilibrium, will be studied in detail.

3.1 Thermodynamics of open systems

To study thermodynamics of an open system, first a defined **control volume** has to be introduced. This control volume is then used for the account of all energy inputs and outputs. An example of such an energy scheme is shown in Figure 3.1.

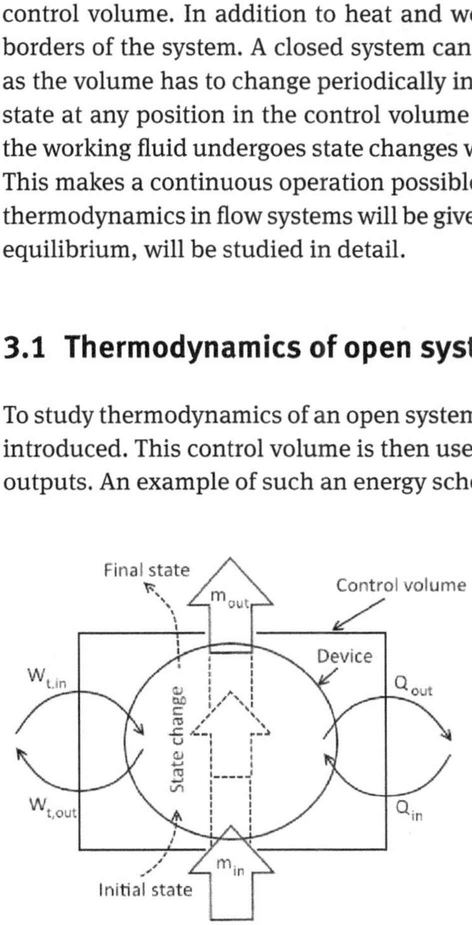

Fig. 3.1: Energy and mass flows transferred across the borders of a control volume.

The energy flows in and out of the control volume then consist of two qualitatively different contributions. First, as in closed systems, there is the transfer of heat and work across the borders of the control volume. The only difference to closed systems is here that a part of the work, the so-called **flow work**, is required to maintain the mass flow through the system. For this reason the **technical work** rather than the total work is transferred freely across the borders of the control volume. This point will be discussed in detail in the following sections. The second part of energy transfers is linked to the mass flow itself. The mass flow usuall exits the system in a thermodynamic state different from the one it had when it entered the system. Thus, the internal energy of matter entering the system may be different from the internal energy of matter leaving the system at the same time, and this difference has to be accounted for in the total energy balance. Furthermore, the matter entering the control volume may have a kinetic or potential energy, varying upon the crossing of the system. It is this mechanical energy carried by the mass flow which underlies the function of wind and water turbines, both of which will be discussed at the end of this chapter. In all other cases considered in this book, however, these two types of energy are nonessential for the functioning of the devices and are thus ignored.

3.1.1 The first law of thermodynamics for open systems

To find a useful expression of the first law of thermodynamics for open systems, a system containing a fixed mass, as shown in Figure 3.2, is considered.

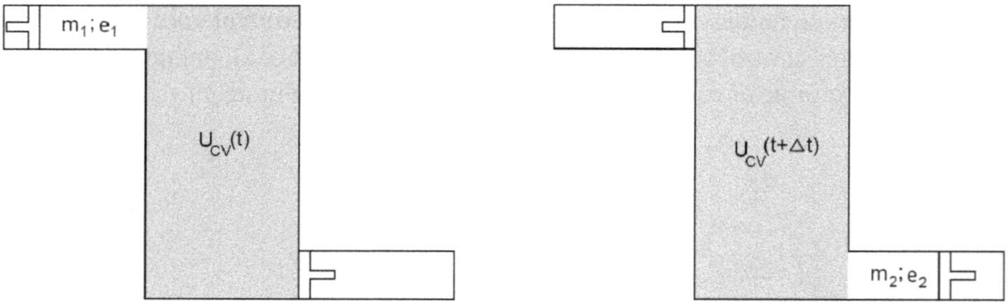

Fig. 3.2: Control volume (*gray area*), moveable masses $m_{1,2}$, and mass specific sum of all energies $e_{1,2}$ of a model system at times t (*left*) and $t + \Delta t$ (*right*).

Initially at time t the greater part of the mass is located in a control volume, and a small mass m_1 is contained in a region labeled '1' directly adjacent to the inlet of the control volume. In the time interval Δt this mass m_1 is pushed entirely into the control volume, while a fraction m_2 of the mass initially contained in the control volume is pushed into a region labeled '2' adjacent to the outlet of the control volume. The whole system, control volume and the volumes taken up by the mass m_1 at time t and the mass m_2 at time $t + \Delta t$, is then a closed system. The mass-specific input and output energies $e_{1,2}$ of the control volume include the mass-specific internal energies $u_{1,2}$, potential energy in the gravitational field $gh_{1,2}$, and kinetic energy $1/2v_{\text{drift};1,2}$ of the finite drift speed of the center of mass m_1 and m_2, respectively. For the further considerations, the kinetic and potential energies are neglected. As already mentioned, they will again play a role in the description of water and wind turbines at the end of this chapter. According to the first law of thermodynamics for closed systems, as given in equation (2.1), this leads to the following expression:

$$\begin{aligned} Q + W = \Delta U = U(t + \Delta t) - U(t) &= U_{CV}(t + \Delta t) + m_2 u_2 - (U_{CV}(t) + m_1 u_1) \\ \Rightarrow U_{CV}(t + \Delta t) - U_{CV}(t) &= Q + W - m_2 u_2 + m_1 u_1 \\ \Rightarrow \frac{U_{CV}(t + \Delta t) - U_{CV}(t)}{\Delta t} &= \frac{1}{\Delta t}(Q + W - m_2 u_2 + m_1 u_1). \end{aligned} \tag{3.1}$$

For $\Delta t \to 0$ the borders of the control volume and the entire system fall together. This also means that the net rates of heat and work transfer of the entire system and of the control volume are identical. From the last line of equation (3.1), the corresponding energy transfer rates can then be obtained:

$$\dot{U}_{CV} = \dot{Q} + \dot{W} - \dot{m}_2 u_2 + \dot{m}_1 u_1 \tag{3.2}$$

A part of the work W is associated with the pressure of the flowing matter at the inlet and outlet. The flow of matter into the system performs work on the system, comparable with a piston pushing mass into the system. Correspondingly, the flow of matter out of the system requires work by the system; the system acts like driving a piston into the surroundings. This **flow work** W_{flow}, is thus performed on the fluid. It is given by

$$W^{flow} = p_1 V_1 - p_2 V_2 = p_1 m_1 v_1 - p_2 m_2 v_2 = \frac{p_1 m_1}{\varrho_1} - \frac{p_2 m_2}{\varrho_2}, \tag{3.3}$$

where v_1 and v_2 are the mass specific volumes of the working fluid at inlet and outlet, respectively, and ϱ_1 and ϱ_2 the corresponding densities. Besides the flow work, the system may provide work to drive some mechanical device, such as shaft work. This remaining part of the work, not associated with the fluid flow, is called **technical work**, W^t:

$$\dot{W} = \dot{W}^t + \dot{W}^{flow} = \dot{W}^t + (\dot{m}_1 p_1 v_1 - \dot{m}_2 p_2 v_2) = \dot{W}^t + \left(\frac{\dot{m}_1 p_1}{\varrho_1} - \frac{\dot{m}_2 p_2}{\varrho_2}\right). \tag{3.4}$$

Substituting equation (3.4) into equation (3.2) the following expression is obtained:

$$\dot{U}_{CV} = \dot{Q} + \dot{W}^t + \dot{m}_1 (u_1 + p_1 v_1) - \dot{m}_2 (u_2 + p_2 v_2) = \dot{Q} + \dot{W}^t + \dot{m}_1 h_1 - \dot{m}_2 h_2, \tag{3.5}$$

with the mass specific enthalpies h_1 and h_2 for the input and output mass flows, respectively. Equation (3.5) is the general formulation of the first law of thermodynamics for an open system with one inlet and one outlet. It is an energy conservation law that accounts for the change in the internal energy of the open system component by all energy transfer processes across its borders.

The internal energy of any control volume in an open system only changes due to the transfer of heat, technical work and the energy balance of the mass flow crossing the system:

$$\dot{U}_{CV} = \dot{Q} + \dot{W}^t + \dot{m}_{in}h_{in} - \dot{m}_{out}h_{out}. \tag{3.6}$$

3.1.2 Steady flow conditions

Apart from start-up or regulation processes, power converters are operated in a steady state characterized by a given mass flow that passes all the system components. This steady state is called the **steady flow equilibrium.** In all cases treated in this book the systems considered are in a steady flow equilibrium, which implies the following properties for any open system component:

1. The sums of the incoming and outgoing mass flow rates are constant in time and identical:
$$\sum \dot{m}_{in} = \sum \dot{m}_{out} = \text{const.} \tag{3.7}$$
2. In a steady state, $\dot{U} = 0$ holds for the control volume and consequently the total rate at which energy is transferred into the control volume equals the total rate at which energy leaves the control volume. This also implies that heat and work powers transferred across the borders of the system component are time independent:
$$\dot{Q} + \dot{W} = \sum \dot{m}_{out}e_{out} - \sum \dot{m}_{in}e_{in}. \tag{3.8}$$
3. All state variables are time independent at every point of the system. Instead, they change as a function of position.

These conditions imply that while mass flows continuously through any fixed control volume, the state of the working fluid at a given point is constant in time. This also means that a mass element m of the working fluid undergoes state changes while passing through the open system which are only dependent on the position inside the system. Thus, the time sequence of state changes of the working fluid in a thermodynamic cycle, as typical for closed systems, is replaced by a sequence of thermodynamic state changes in space.

In the steady flow equilibrium, $\dot{m}_1 = \dot{m}_2 =: \dot{m}$ holds for an inlet at state '1' and an outlet at state '2', as apparent from equation (3.7). This means that for open systems in a steady flow equilibrium it is convenient to calculate mass specific values for all state variables as well as for the transferred heat and work. These values multiplied with the mass current $\dot{m}$ then directly yield the respective powers transferred across the borders of the system. Thus, the first law of thermodynamics for an open system component given in equation (3.5) can be simplified for the steady flow equilibrium:

In a steady flow equilibrium the internal energy of any control volume is constant in time. The amounts of work and heat transferred across the borders of the control volume are identical to the energy balance of the mass flow:

!

$$\begin{aligned} &\dot{m}(h_{\text{out}} - h_{\text{in}}) = \dot{m}\Delta h = \dot{Q} + \dot{W}^{\text{t}} \\ \Leftrightarrow\quad &\Delta h = q + w^{\text{t}} \\ \Leftrightarrow\quad &dh = \delta q + \delta w^{\text{t}}. \end{aligned} \tag{3.9}$$

Comparing equation (3.9) with the first law of thermodynamics for closed systems given in equation (2.1), one observes that both formulations of the first law of thermodynamics are very similar. The difference is that in open systems in a steady flow equilibrium the internal energy is replaced by the enthalpy accounting for the flow work. In doing so the usable work, namely the technical work, crossing the borders of the system is obtained on the right-hand side.

3.1.3 Technical work

In this section the mathematical expression for the technical work, i.e. the part of the work crossing the borders of the system component that can be used externally, is derived for a system component in a steady flow equilibrium. It can directly be read out from the total differential of the mass specific enthalpy, which reads (for details see Section A.1 of the appendix):

$$dh = d(u + pv) = du + pdv + vdp = Tds - pdv + pdv + vdp = Tds + vdp. \tag{3.10}$$

Here the term Tds can be identified with the reversibly exchanged heat δq. A comparison of equation (3.9) with equation (3.10) then leads to the following term for the technical work w^{t}:

$$\delta w^{\text{t}} = vdp = 1/\varrho\, dp. \tag{3.11}$$

This means that technical work is crossing the borders of an open system component in a steady flow equilibrium only if the working fluid undergoes a change in pressure while running through the component. To illustrate this, in Figure 3.3 a model process where the working fluid undergoes such a state change while moving through an open component is demonstrated with the example of a periodically fed piston compressor.

In the first step, a mass m is fed to the compressor at the inlet pressure $p = p_1$. The system is then in the defined state '1' characterized by pressure p_1 and volume V_1. To establish this state the system actively pushes the moveable border which means, according to the sign convention used, that a negative work $-p_1 V_1$ is transferred across the borders of the system in this step. In the second step the mass in the closed compressor is compressed until the pressure $p = p_2$ is reached, a process accompanied by a positive work W transferred into the system. Thereby the volume of the system is

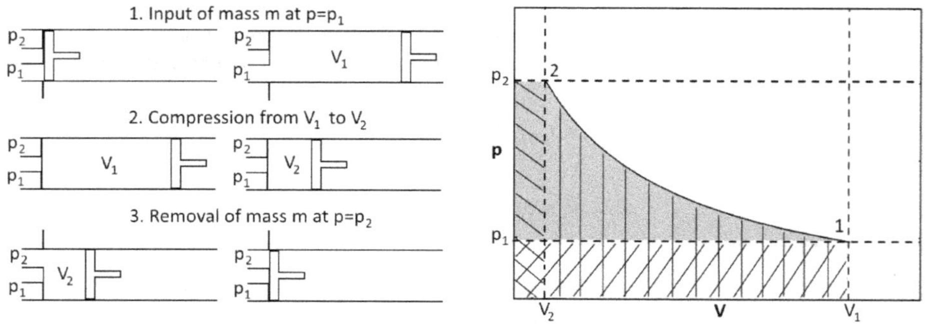

Fig. 3.3: Example process for an open component changing the state of the working fluid; *Left:* Sketch of the process steps: 1. injection of mass m at the inlet at $p = p_1$; 2. compression within the control volume; 3. removal of mass m at the outlet at $p = p_2$. *Right:* pV-diagram, the work corresponding to the three steps is marked as slanted right (step 1), vertical (step 2), and slanted left (step 3); the technical work is marked in *gray*.

reduced to $V = V_2$. Finally, in a third step the mass m is pressed out of the compressor through the outlet at $p = p_2$, which again requires a positive external work $p_2 V_2$. The overall technical work transferred across the borders of the system in this cyclic process is then given by

$$W^{\mathrm{t}} = -p_1 V_1 + W + p_2 V_2 = -\int_1^2 pdV - (p_1 V_1 - p_2 V_2) = W - W^{\mathrm{flow}}. \tag{3.12}$$

In Figure 3.3 the areas marked *slanted-left lines* and *vertical lines* count positively, as work is performed at the system in the corresponding steps 2 and 3, while the area marked *slanted-right lines*, conversely, counts negatively as the system performs work against the ambient state there. The technical work thus corresponds to the sum of the areas only marked *vertical lines* or *slanted-left lines* but not '*slanted-right lines*'. This area can then indeed be calculated by an integration along the pressure axis:

$$W^{\mathrm{t}} = \int_1^2 Vdp, \tag{3.13}$$

or in the mass-specific form

$$w^{\mathrm{t}} = \int_1^2 vdp = \int_1^2 \frac{1}{\varrho} dp. \tag{3.14}$$

A negative sign means that the system performs technical work, which is again in accordance with the sign convention. If a continuous compressor is used instead of the piston compressor the integral from state '1' to state '2' can be interpreted as a spatial integral from the inlet of an open-system component to its outlet.

3.2 Exergy of open systems

To define the exergy of an open system, the states of the mass flows entering and leaving the system have to be considered. The modified definition for the mass specific exergy ψ of an open system reads as follows:

The exergy per unit mass ψ_i of a working fluid which is entering an open system in the state 'i' is the maximal amount of technical work extractable from the open system when at the exit the working fluid is in equilibrium with the ambient state.

This means that for a working fluid in a given state 'i' an open system component is envisioned whose outlets are in equilibrium with the ambient state. Note that the definition of the exergy includes again kinetic and potential energies of the working fluid in state 'i'. However, as both forms of energy are pure exergy, they are not considered in the formulas below. If necessary they can be taken into account by simply adding them to the right-hand side in equation (3.15). The extractable work is path dependent and, as in the case of closed systems considered in Chapter 2, the optimal path consists of first an isentropic expansion until $T = T_{\text{am}}$ is reached with a subsequent isothermal heat transfer at $T = T_{\text{am}}$. With the first law of thermodynamics in its formulation for open systems in a steady flow equilibrium in equation (3.9) the maximal extractable mass-specific technical work against the ambient state ${}_{\text{i}}w^{\text{t}}_{\text{am}}$ is then given by

$$\begin{aligned} {}_{\text{i}}w^{\text{t}}_{\text{am}} &= (h_{\text{am}} - h_{\text{i}}) - {}_{\text{i}}q_{\text{am}} = (h_{\text{am}} - h_{\text{i}}) - T_{\text{am}}(s_{\text{am}} - s_{\text{i}}) \\ \Rightarrow \quad \psi_{\text{i}} &= -\,{}_{\text{i}}w^{\text{t}}_{\text{am}} = (h_{\text{i}} - h_{\text{am}}) - T_{\text{am}}(s_{\text{i}} - s_{\text{am}}) \end{aligned} \tag{3.15}$$

The expression for the exergy of an open system, in contrast to the corresponding expression for a closed system given in equation (2.11), does not contain an additional term for the work of displacement w^{dis}, i.e. work performed against the ambient state. The reason for this is that the work connected with the mass flow through the system is already taken into account in the definition of the technical work w^{t} via the flow work w^{flow}. As exergy differences rather than exergies are often of interest, the exergy function is defined analogously to the corresponding quantity for closed systems given in equation (2.12):

$$\psi^{\text{x}}_{\text{i}} = \psi_{\text{i}} + h_{\text{am}} - T_{\text{am}}s_{\text{am}} = h_{\text{i}} - T_{\text{am}}s_{\text{i}}. \tag{3.16}$$

Differences in exergy functions and differences in exergies are again identical. The exergy content of energy transferred across the borders of an open system can be interpreted in a similar way, as laid out in detail in Section 2.3 for closed systems. As the work of displacement does not play a role anymore, the definition of the anergy content of energy transferred to an open system simplifies:

The anergy content of energy given to an open system is the entropy free part of the energy. It corresponds to the minimal amount of heat that has to be rejected in order to prevent an accumulation of entropy, i.e. $A = T_{am} \Delta S$.

With this the exergy contents $\Psi({}_1W_2)$ and $\Psi({}_1Q_{2,h})$ of technical work and heat transferred isothermally at $T = T_h$, respectively, can be given

$$\begin{aligned} \Psi({}_1W_2) &= {}_1W_2^t \\ \Psi({}_1Q_{2,h}) &= {}_1Q_2 - \frac{T_{am}\Delta S}{{}_1Q_2} = {}_1Q_2 \cdot \left(1 - \frac{T_{am}}{T_h}\right). \end{aligned} \tag{3.17}$$

Here, the uppercase letters indicate that absolute values rather than mass specific values are given.

3.3 Important example components

In this section examples of important open systems are presented and their second-law efficiency is given. The heat exchanger is an important component found in many energy conversion systems, while water and wind turbines are instructive examples and, beyond that, most important for renewable energy production.

3.3.1 Heat exchangers

To transfer heat from one fluid to another one without mixing the respective streams, closed heat exchangers are used. A simple illustration is depicted in Figure 3.4, where hot source and cold target fluids stream through two tubes which are both isolated against the surroundings but in close thermal contact with each other to allow for an efficient heat transfer between the tubes. The borders of the thermodynamic system can be thought of as being around the insulating walls with two inlets and two outlets, one of each for the relatively hot source system and the relatively cold target system, respectively.

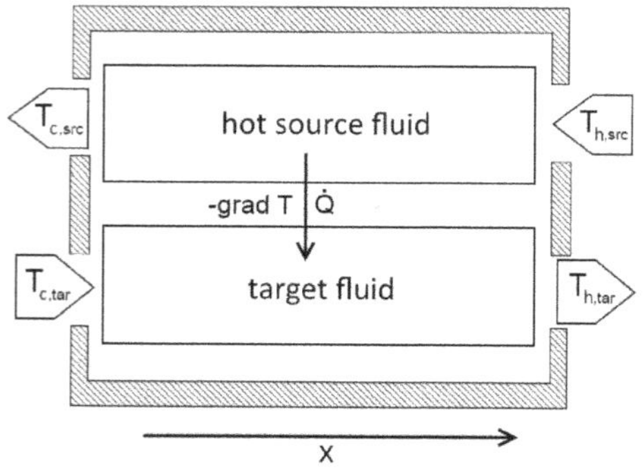

Fig. 3.4: A counterflow heat exchanger where the source system (top) cools down from $T_{h,src}$ to $T_{c,src}$ while providing heat to the target system (bottom) heating it from $T_{c,tar}$ to $T_{h,tar}$. The counterflow keeps variations in the temperature gradient between the two systems minimal.

In an ideal heat exchanger no energy is lost to the surroundings, i.e. the entire heat which leaves the source system is transferred to the target system. Thus the first law efficiency is $\eta_1 = 1$. The heat is transferred between the two systems via heat conduction through the wall, which is driven by the temperature gradient:

$$\dot{q}(x) \propto -\nabla T(x). \tag{3.18}$$

Hence, clearly a higher power throughput can be achieved by a higher temperature gradient. However, due to the temperature gradient the heat is rejected from the source system at a different temperature than it enters the target system. For this reason, the exergy content of the heat diminishes (see Section 2.2), leading to a second-law efficiency of $\eta_2 < 1$ for any closed heat exchanger. The sketch in Figure 3.4 depicts the most common type of heat exchanger, the counterflow heat exchanger, where the hot and cold ends of both streams are each on the same side. This configuration leads to the lowest exergy losses of all closed heat exchangers.

To calculate the second-law efficiency of such a counterflow heat exchanger, note first that the heat transfer is an isobaric process for both systems, which means that no technical work is transferred during the process. The second-law efficiency of the heat exchanger is then the ratio of the exergy gained by the working fluid of the target system and the exergy lost by the hot source stream:

$$\begin{aligned}\eta_2 &= \frac{\Delta \dot{\Psi}_{\text{tar}}}{\Delta \dot{\Psi}_{\text{src}}} = \frac{\dot{m}_{\text{tar}} |\psi_{\text{h,tar}} - \psi_{\text{c,tar}}|}{\dot{m}_{\text{src}} |\psi_{\text{h,src}} - \psi_{\text{c,src}}|} \\ &= \frac{\dot{m}_{\text{tar}}}{\dot{m}_{\text{src}}} \cdot \frac{(h_{\text{h,tar}} - h_{\text{c,tar}}) - T_{\text{am}}(s_{\text{h,tar}} - s_{\text{c,tar}})}{(h_{\text{h,src}} - h_{\text{c,src}}) - T_{\text{am}}(s_{\text{h,src}} - s_{\text{c,src}})},\end{aligned} \tag{3.19}$$

where the index 'tar' stands for quantities in the target system and the index 'src' for quantities in the source system. In equation (3.19) the states at the inlets and outlets of both systems are used. No further assumptions have to be made. Whenever tabulated values for all thermodynamic quantities used in equation (3.19) are accessible, it is the best choice for the calculation of the second-law efficiency.

Another way of calculating the second-law efficiency of the heat exchanger is to consider the transferred energy itself, i.e. to divide the exergy content of the heat entering the target system by the exergy content of the heat rejected by the source system. As the temperature changes during an isobaric heating process and heat and exergy transfers are temperature dependent, an integration over the temperature range considered has to be performed to calculate the total mass specific exergy $\psi({}_1q_2)$ transferred into an open system between the temperatures T_1 and T_2 in an isobaric process. In the following, first any phase change of the working fluid is excluded:

$$\psi({}_1q_2) = \int_1^2 d\psi = \int_1^2 \left(1 - \frac{T_{\text{am}}}{T}\right) \delta q = \int_{T_1}^{T_2} c_{\text{p}}(T) \cdot \left(1 - \frac{T_{\text{am}}}{T}\right) dT. \tag{3.20}$$

Here $c_p(T)$ is the isobaric heat capacity. As c_p is positive, this leads to a positive exergy transfer when the system is heated and a negative exergy transfer when the system is cooled. In the case of a constant heat capacity $c_p(T) \to c_p$ the integration in equation (3.20) can be performed, yielding

$$\psi({}_1q_2) = c_p\left((T_2 - T_1) - T_{am} \cdot \ln\left(\frac{T_2}{T_1}\right)\right). \tag{3.21}$$

Using equation (3.20), the second-law efficiency for a heat exchanger can thus be calculated in terms of the exergies transferred during the heat exchange process:

$$\begin{aligned}\eta_2 &= \frac{\dot{m}_{tar}}{\dot{m}_{src}} \cdot \left|\frac{\psi({}_1q_{2,tar}^{isobar})}{\psi({}_1q_{2,src}^{isobar})}\right| = \frac{\dot{m}_{tar}}{\dot{m}_{src}} \cdot \left|\frac{\int_{c,tar}^{h,tar} d\psi_{tar}}{\int_{h,src}^{c,src} d\psi_{src}}\right| \\ &= \frac{\dot{m}_{tar}}{\dot{m}_{src}} \cdot \frac{\int_{T_{c,tar}}^{T_{h,tar}} c_{p,tar}(T)\,(1 - T_{am}/T)\,dT}{\int_{T_{c,src}}^{T_{h,src}} c_{p,src}(T)\,(1 - T_{am}/T)\,dT}\end{aligned} \tag{3.22}$$

In a steady-flow equilibrium, the amount of heat transferred between both systems is time independent. This has implications for the ratio between both mass flows. At a steady-flow equilibrium, the energy rate balance reduces to

$$\begin{aligned}&\dot{m}_{src}(h_{h,src} - h_{c,src}) = \dot{m}_{tar}(h_{h,tar} - h_{c,tar}) \\ \Leftrightarrow\ &\dot{m}_{src} \int_{T_{c,src}}^{T_{h,src}} c_{p,src}(T)dT = \dot{m}_{tar} \int_{T_{c,tar}}^{T_{h,tar}} c_{p,tar}(T)dT.\end{aligned} \tag{3.23}$$

Assuming constant, but not necessarily identical heat capacitances $c_{p,src}$ and $c_{p,tar}$, the last expression can be further simplified:

$$\begin{aligned}&\dot{m}_{src}c_{p,src}(T_{h,src} - T_{c,src}) = \dot{m}_{tar}c_{p,tar}(T_{h,tar} - T_{c,tar}) \\ \Rightarrow\ &\frac{\dot{m}_{src}}{\dot{m}_{tar}} = \frac{c_{p,tar}(T_{h,tar} - T_{c,tar})}{c_{p,src}(T_{h,src} - T_{c,src})} =: \frac{c_{p,tar}\Delta T_{tar}}{c_{p,src}\Delta T_{src}}.\end{aligned} \tag{3.24}$$

The second-law efficiency η_2 of the counterflow heat exchanger then reads:

$$\eta_2 = \frac{1 - T_{am}/\Delta T_{tar} \cdot \ln(T_{h,tar}/T_{c,tar})}{1 - T_{am}/\Delta T_{src} \cdot \ln(T_{h,src}/T_{c,src})}. \tag{3.25}$$

When phase transitions of the working fluids are occurring in the source or the target system, as relevant for steam power plants as described in Chapter 4, equation (3.22) has to be written in a more general form:

$$\begin{aligned}\eta_2 &= \frac{\dot{m}_{tar}}{\dot{m}_{src}} \cdot \left|\frac{\psi({}_1q_{2,tar}^{isobar})}{\psi({}_1q_{2,src}^{isobar})}\right| \\ &= \frac{\dot{m}_{tar}}{\dot{m}_{src}} \cdot \frac{(x_{2,tar} - x_{1,tar}) \cdot q_{l,tar}\left(1 - T_{am}/T_{pc,tar}\right) + \int_{T_{1,tar}}^{T_{2,tar}} c_{p,tar}(T)\,(1 - T_{am}/T)\,dT}{(x_{1,src} - x_{2,src}) \cdot q_{l,src}\left(1 - T_{am}/T_{pc,src}\right) + \int_{T_{2,src}}^{T_{1,src}} c_{p,src}(T)\,(1 - T_{am}/T)\,dT},\end{aligned} \tag{3.26}$$

where $x_{\mathrm{src;tar}}$ stands for the steam mass fractions, $q_{\mathrm{l,src;tar}}$ for the latent heats, and $T_{\mathrm{pc,src;tar}}$ for the temperature of the phase change for the respective media. For details concerning phase transitions see Section A.4 of the appendix.

3.3.2 Water turbines

In a water turbine a high pressure is generated by a water column standing on the turbine inlet. As water is incompressible, the pressure is directly linked to the height h of the water column:

$$p_{\mathrm{in}} = \varrho g h + p_{\mathrm{out}}. \tag{3.27}$$

Here ϱ is the volume density of the water, p_{out} the pressure at the turbine outlet, and g the gravity acceleration of the earth. Inside the turbine no density changes occur in the incompressible water, and also no heat transfer occurs, as an ideal process is assumed. As the process is isochoric, no volume work is crossing the borders of the system, and from equation (3.4) one can conclude that the flow work is equal to the negative technical work. Thus, the energy source provides the flow work which can then be fully transformed into technical work by the turbine:

$$w^{\mathrm{t}} = (p_{\mathrm{out}} - p_{\mathrm{in}})v = (p_{\mathrm{out}} - (\varrho g h + p_{\mathrm{out}}))v = -\frac{\varrho g h}{\varrho} = -gh. \tag{3.28}$$

This means that the full potential energy $E_{\mathrm{pot}} = mgh$ of a mass m of water added at the top of the water column is transferred out of the system as technical work. As stated above, potential energy in the gravitational field is hence pure exergy. As both the input and output energies are pure exergy, the first- and second-law efficiencies of a water turbine are identical and only determined by mechanical losses.

3.3.3 Wind turbines

The energy fed to an open system component can be mechanical but not linked to changes in the internal energy of the working medium. This is realized in wind turbines, where the kinetic energy of the moving air field is utilized to produce electrical energy. The kinetic energy is pure exergy and has to be added to the total internal energies per volume at the inlet and outlet of the system. In the case of a wind turbine, the pressure and consequently the density of the air before and after the turbine is identical, as they are subject to essentially the same surroundings. The only change occurs in the kinetic energy of the moving air field, which is significantly lower at the outlet than the inlet. While the kinetic energy is pure exergy, it still cannot be fully used, as the air has to continuously pass the turbine and hence has to retain a certain speed and kinetic energy. Under the assumption of a continuous air flow with constant density the expression

$$\dot{m} = \varrho \dot{V} = \varrho \vec{A} \cdot \vec{v} =: A v_{\perp} \tag{3.29}$$

holds, where $\vec{v}$ is the velocity of the wind field, $\vec{A} = A\vec{n}$ the vector area crossed by the wind field, and $v_\perp$ the velocity component perpendicular to the area. For simplicity in the further discussion, a perpendicular flow through all relevant areas is assumed, which means that $v = v_\perp$ holds for all velocities, and only scalar quantities have to be considered.

A reduction of the speed of the air field leads to an increase of the cross-section area A according to the continuity equation

$$\dot{V} = A_{\text{in}} \cdot v_{\text{in}} = A_{\text{turb}} \cdot v_{\text{turb}} = A_{\text{out}} \cdot v_{\text{out}}. \tag{3.30}$$

Here, the air speed $v_{\text{turb}} = 1/2(v_{\text{in}} + v_{\text{out}})$ and consequently the cross-sectional area $A_{\text{turb}} = 1/2(A_{\text{in}} + A_{\text{out}})$ of the wind turbine are approximated by the arithmetic average of the corresponding input and output values, as shown in Figure 3.5.

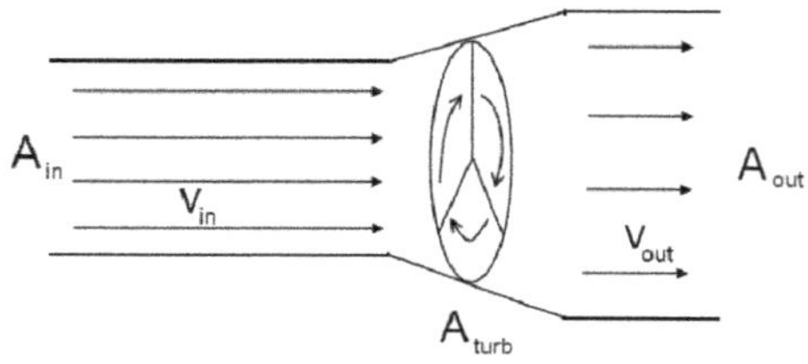

Fig. 3.5: Schematic of the air flow field and its change due to a wind turbine.

The total power provided by the moving air field to the wind turbine at a given wind velocity v_{in} and turbine area A_{turb} can then be written as

$$\begin{aligned} P &= \frac{1}{2}\dot{m}(v_{\text{in}}^2 - v_{\text{out}}^2) = \frac{1}{2}\varrho\left(A_{\text{in}} \cdot v_{\text{in}}^3 - A_{\text{out}} \cdot v_{\text{out}}^3\right) \\ &= \frac{1}{2}\varrho A_{\text{turb}} \cdot v_{\text{turb}}\left(v_{\text{in}}^2 - v_{\text{out}}^2\right) = \frac{1}{4}\varrho A_{\text{turb}}\left(v_{\text{in}} + v_{\text{out}}\right)\left(v_{\text{in}}^2 - v_{\text{out}}^2\right) \\ &= \frac{1}{4}\varrho A_{\text{turb}}\left(v_{\text{in}}^3 + v_{\text{in}}^2 v_{\text{out}} - v_{\text{in}} v_{\text{out}}^2 - v_{\text{out}}^3\right). \end{aligned} \tag{3.31}$$

At the optimal choice of v_{out} this power is maximal:

$$\begin{aligned} \frac{dP}{dv_{\text{out}}} &= \frac{1}{4}\varrho A_{\text{turb}}\left(-3v_{\text{out}}^2 - 2v_{\text{in}}v_{\text{out}} + v_{\text{in}}^2\right) \stackrel{!}{=} 0 \\ \Rightarrow \quad & -3v_{\text{out}}^2 - 2v_{\text{in}}v_{\text{out}} + v_{\text{in}}^2 = 0 \\ \Rightarrow \quad & v_{\text{out}} = \frac{1}{3}v_{\text{in}} \end{aligned} \tag{3.32}$$

The maximal second-law efficiency of the wind turbine is then given by the comparison of this maximal extracted power with the total kinetic power of the wind field passing the area of the wind turbine unperturbed:

$$\eta = \frac{1/4\varrho A_{\text{turb}}(32/27)v_{\text{in}}^3}{1/2\varrho A_{\text{turb}}v_{\text{in}}^3} = \frac{16}{27} = 59\,\%. \tag{3.33}$$

Under the assumptions made, this is the maximum first- and second-law efficiency of a wind turbine. This efficiency is called **Betz efficiency** after the German Physicist Albert Betz.

4 Thermal power plants

In this chapter different designs of conventional thermal power plants and the corresponding thermodynamic cycles are discussed in detail. Special emphasis is placed on exergy efficiency considerations of the individual components and the overall power plants. The two main types of thermal power plants are **steam power plants** and **gas turbine power plants.** In a steam power plant the working fluid changes its phase during the operation of the power plant. Steam power plants are by far the most common type of power generators accounting for around 60 % (2012) of the electrical final energy output (based on IEA data from "Key World Energy Statistics 2014" © OECD/IEA 2014, IEA Publishing; modified by the authors Licence: http://www.iea.org/t&c/termsandconditions/). Coal-, oil-, and biomass-fueled power plants are of this type, as are also nuclear and solar thermal power plants. Gas turbine power plants have become quite common today (23 % in 2012 worldwide again according to the Key World Energy Statistics from the IEA (2014)) and may further gain in importance in the future for two major reasons. First, for the production of electrical energy, a modern coal-fired power plant typically releases $\approx$ 900 g/kWh of carbon dioxide, whereas a gas turbine power plant typically only releases $\approx$ 600 g/kWh CO_2. The reason for this is the higher heating value of methane compared to coal. The second important reason for the rising importance of gas turbine power plants is the fact that their power output can be regulated very quickly, on a timescale of minutes, which gains in importance as the share of fluctuating renewable energies in the overall electricity production rises. Both types of thermal power plants can be combined to form the so-called **combined cycle power plant** (CCPP), which is fed by natural gas and reaches the highest efficiencies of all thermal power plants. Such a power plant can reach first law efficiencies of about 60 % and CO_2 emission values as low as 350 g/kWh.

4.1 Steam power plants

A steam power plant can be separated into three different working units, as shown in a schematic in Figure 4.1.

The central part is a heat engine, where heat is converted first into mechanical, technical work, and then into electrical energy. This is done by relaxing the pressurized heated working fluid, typically water, across a turbine which is mounted on a joint shaft with an electrical generator. The heat is provided to the heat engine by some heated medium through a closed heat exchanger. For different types of power plants the generation of the heat carried by this heated medium is achieved in different ways. If the primary energy carrier is a chemical fuel such as coal or biomass, heat generation is achieved by the combustion of this fuel, and the heated medium is the

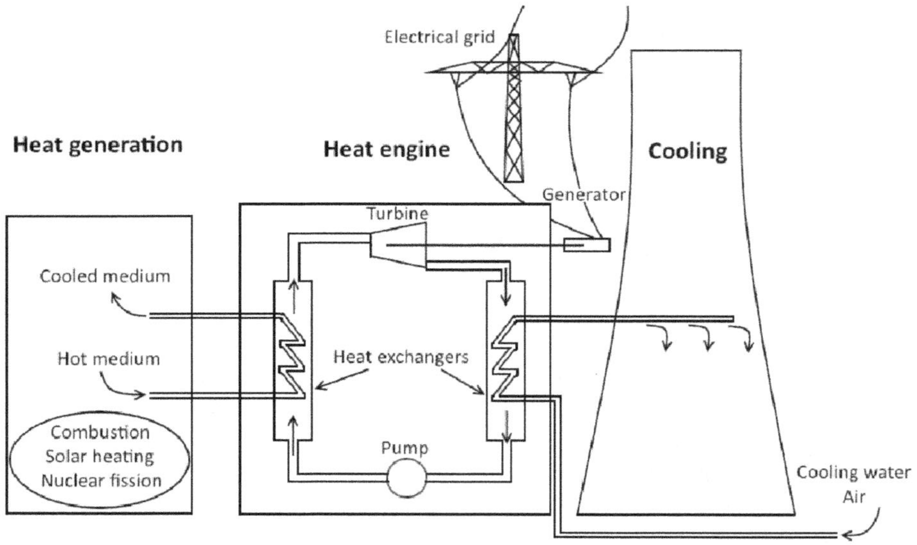

Fig. 4.1: Schematic of a steam power plant.

hot flue gas of the combustion process. In a solar thermal power plant, typically a heat transfer oil or air is heated by focused solar radiation, and in a nuclear power plant water is heated in a reactor vessel containing the nuclear fuel rods. The cooling of all thermal power plants is typically achieved by cooling water, which is continuously extracted from some source such as a river and evaporated into a cooling tower. Especially for solar thermal power plants, air is also sometimes used, as in desert regions, where these plants are working best, water is a scarce resource.

4.1.1 Rankine cycle

In this section, a closer look at the central heat engine part of the power plant is taken. The thermodynamic cycle capturing the succession of the processes occurring in the heat engine is the **Rankine cycle.** In this cycle the working fluid, typically water, runs through the basic components shown in Figure 4.2.

The working fluid changes its phase during the Rankine cycle. A short overview of the relevant properties of phase transitions and the nomenclature is given in the appendix in Section A.4.

The Rankine cycle operates between two pressure levels $p = p_{2,3}$ and $p = p_{4,1}$ and is composed of two isobaric and two adiabatic processes. Both heat exchange processes are isobaric processes at the respective pressure levels, which means that the technical work is equal to zero. Conversely, the two processes where the pressure is changed,

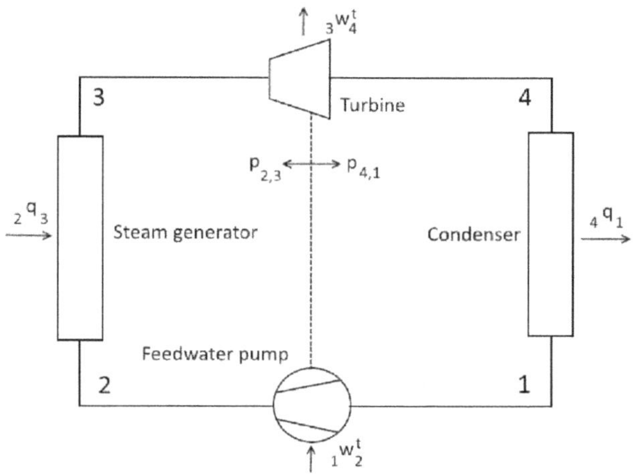

Fig. 4.2: Schematic of the heat engine part realized in a steam power plant. The *vertical dashed line* separates the two pressure levels in the cycle.

the compression and relaxation of the working fluid, are adiabatic, i.e. $q = 0$ holds for both. It is thus comparatively easy to analyze the cycle, as for all four state changes either the transferred heat vanishes or the transferred technical work. In the following the four thermodynamic processes are described in detail. The numbers refer to the states of the working fluid at the inlet and outlet of the different components as indicated in Figure 4.2.

1 ⟶ 2: Feed water is compressed adiabatically in the liquid phase from the lower to the higher pressure level by the **feed water pump**. This compression does not require a large amount of energy, because water has a comparatively high density in its liquid phase, minimizing the technical work necessary for this process. The total mass specific technical work $_1w^t_2$ needed in this step is

$$_1w^t_2 = \int_1^2 v dp \approx v_1(p_2 - p_1) = (1/\varrho_1)(p_2 - p_1) > 0. \tag{4.1}$$

2 ⟶ 3: The water is fed into the **steam generator** by the feed water pump, where it takes up the mass specific heat $_2q_3$, which first brings it to its boiling point and then fully evaporates it, both isobarically at the high pressure level. The heat $_2q_3$ is provided by the hot combustion products of the fuel and can be calculated via

$$_2q_3 = h_3 - h_2 \approx c_p \cdot (T_3 - T_2) + q_{l,p=p_2} > 0, \tag{4.2}$$

where c_p is the mass specific heat capacity of the fluid in its liquid phase, which is assumed to be constant and $q_{l,p=p_2}$ the latent heat necessary for the evaporation of the fluid at the pressure level $p = p_2$.

3 ⟶ 4: The fully evaporated working fluid at the higher pressure level is fed to the **turbine**, where it is expanded adiabatically to the lower pressure level. During this process the turbine provides the technical work ${}_3w_4^t$ per unit mass of the fluid to the outside. The vapor is partly converted back into liquid fluid during the process, i.e. point 4 lies in the region of coexisting phases. According to the first law of thermodynamics for open systems, equation (3.9), ${}_3w_4^t$ is given by

$$ {}_3w_4^t = h_4 - h_3 = \int_3^4 v(p)dp = \int_3^4 \frac{1}{\varrho(p)}dp < 0. \tag{4.3}$$

4 ⟶ 1: The liquid/vapor fluid mixture of the working fluid runs through the **condenser.** In this part the steam mass fraction x_4 of the fluid condenses isobarically and the fluid rejects the mass specific waste heat ${}_4q_1$ to the environment:

$$ {}_4q_1 = h_1 - h_4 = -x_4 \cdot q_{\mathrm{lat},p=p_1} < 0. \tag{4.4}$$

Here, $q_{\mathrm{lat},p=p_1}$ stands for the mass specific latent heat of the thermal fluid at the lower pressure level $p = p_1$. Isobaric processes are not associated with the transfer of technical work, as can be seen from equation (3.11). The condensation process takes place in the region of coexisting phases, which means that the process is also isothermal. The temperature is typically only slightly above the ambient temperature to minimize exergy losses in this step.

The Ts- and pv-diagrams of the ideal Rankine cycle are shown in Figure 4.3. In an ideal process all steps are assumed to be reversible, i.e. infinitely slow.

The overall first law efficiency of the ideal Rankine cycle can readily be given using the first law of thermodynamics in its form for open systems in a steady flow equilib-

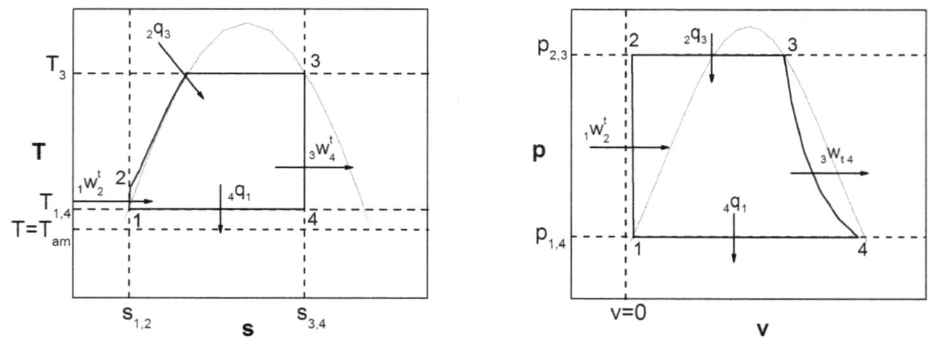

Fig. 4.3: Ts- and pv-diagrams of the ideal Rankine cycle consisting of two isobars realized in the steam generator (2 ⟶ 3) and condenser (4 ⟶ 1) linked by two adiabatic state changes realized in the turbine (3 ⟶ 4) and feed water pump (1 ⟶ 2). The region of coexisting phases is below the *gray curve* in both diagrams.

rium, as given in equation (3.9):

$$\eta_1 = \frac{-(\overbrace{{}_1w_2^t}^{>0} + \overbrace{{}_3w_4^t}^{<0})}{{}_2q_3} = \frac{(h_1 - h_2) + (h_3 - h_4)}{(h_3 - h_2)} = 1 - \frac{h_4 - h_1}{h_3 - h_2} \tag{4.5}$$
$$= 1 - \frac{-{}_4q_1}{{}_2q_3} = 1 - \frac{|q_{\text{out}}|}{|q_{\text{in}}|}.$$

η_1 can be further evaluated with the following definition of an average temperature in the steam generator:

$${}_2q_3 = \int_{s_2}^{s_3} T(s)ds =: \overline{T}_{\text{in}}(s_3 - s_2). \tag{4.6}$$

$\overline{T}_{\text{in}}$ thus marks the height of a rectangle with a width spanning from s_2 to s_3 and an area identical to the area below the $T(s)$-curve between points 2 and 3 in the Ts-diagram in Figure 4.3. Together with equation (4.5) and $s_3 = s_4$ and $s_1 = s_2$, as evident from Figure 4.3, this leads to the following expression for the first law efficiency of the ideal Rankine cycle:

$$\eta_1 = 1 - \frac{|q_{\text{out}}|}{|q_{\text{in}}|} = 1 - \frac{T_{\text{out}}(s_4 - s_1)}{\overline{T}_{\text{in}}(s_3 - s_2)} = 1 - \frac{T_{\text{out}}}{\overline{T}_{\text{in}}}. \tag{4.7}$$

It follows directly from equation (4.7) that for an efficient process the temperature of the heat input $\overline{T}_{\text{in}}$ should be chosen as high as possible and the temperature T_{out} for the waste heat output as low as possible. The choice for the latter is limited downwards by the ambient temperature. In this limiting case the maximal possible first-law efficiency of a thermal engine following the Rankine cycle with a given $\overline{T}_{\text{in}}$ is then given. It can be directly identified with the exergy content of the driving heat ${}_2q_3$, as described in Section 3.2:

$$\eta_{1,\max} = \frac{\psi({}_2q_3)}{{}_2q_3} = \frac{{}_2q_3 - T_{\text{am}}\Delta s}{\overline{T}_{\text{in}}\Delta s} = 1 - \frac{T_{\text{am}}}{\overline{T}_{\text{in}}}. \tag{4.8}$$

An ideal Rankine cycle thus has a second-law efficiency of $\eta_2 = 1$, provided that the lower temperature level is chosen at $T = T_{\text{am}}$. Although the details of the thermodynamic cycles in real power plants differ from the ideal Rankine cycle, the expression for the first law efficiency given in equation (4.7) is a good estimate for real power plants. It also has implications for the two pressure levels of the process as temperature and pressure levels are linked through the phase transitions during evaporation and condensation of the working fluid. The upper pressure level should be as high as possible to also get the temperature $\overline{T}_{\text{in}}$ as high as possible. For the lower pressure level there seem to be at first sight two canonical choices. Either one chooses the atmospheric pressure $p = p_{\text{am}}$ or a lower pressure $p < p_{\text{am}}$ such that the working fluid leaves the condenser at $T = T_{\text{am}}$, which is the boiling point at this pressure. For water to undergo the phase transition at T_{am}, the lower pressure level has to be chosen at ca. $p_1 = 20$ mbar. However, according to equation (4.7) the latter choice of parameters leads to the higher first-law efficiency. This is the reason why a condenser is used in steam power plants. The feed water cycle is then closed, and the pressure in the condenser can be fixed at a desired value. The alternative would be to continuously feed the feed water pump with fresh water and release the steam/liquid mixture at the turbine outlet to the environment. The lower pressure level would then be fixed at $p = p_{am}$, leading to a lower overall first-law efficiency.

In Figure 4.4 an ideal Rankine cycle is contrasted with an ideal Carnot cycle in the region of coexisting phases.

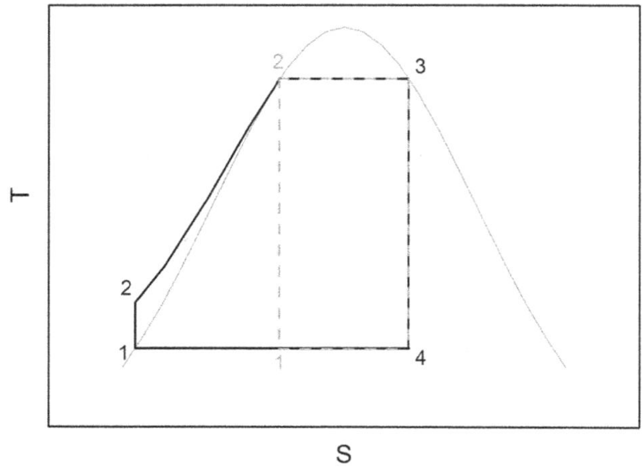

Fig. 4.4: Ts-diagram of an ideal Rankine cycles (*thick solid line*) and a comparable, ideal Carnot cycle in the region of coexisting phases (*dashed line*).

The maximal first-law efficiency of the Rankine cycle is lower than the first-law efficiency of the Carnot cycle. The reason for the lower efficiency is that the input of the heat ${}_2q_3$ is at a lower average temperature as the temperature changes during the process:

$$\begin{aligned} &\overline{T}_{in} < T_3 \\ \Rightarrow \quad &\eta_{1,\text{Rankine}} = 1 - \frac{T_1}{\overline{T}_{in}} < 1 - \frac{T_1}{T_3} = \eta_{1,\text{Carnot}}. \end{aligned} \tag{4.9}$$

There are two main reasons why in a real steam power plant a Rankine cycle and not the principally more efficient Carnot cycle is realized.

1. The source of the heat provided to the working fluid are hot combustion gases in coal or biomass fueled power plants, or another heated medium in the case of nuclear and solar thermal power plants. This thermal source has to be cooled down as far as possible to extract as much energy as possible from it. In the Carnot process the heat source must be at a temperature $T \geq T_3$, while in the Rankine process heat provided at temperatures between T_2 and T_3 can also be used.

2. In the Rankine cycle the feed water pump (process $1 \rightarrow 2$) is compressing liquid water only, whereas in the case of the Carnot process it is operating in the region of coexisting phases. While the former case is easy to handle for suitable pumps, it is considerably more difficult to operate a pump in the region of coexisting phases, making the Carnot cycle an undesirable choice in terms of pump stability and efficiency.

4.1.2 Efficiency of a steam power plant

In this section the overall first- and second-law efficiencies of a steam power plant for the conversion of the energy input into electrical energy will be discussed in detail. Therefore the energy conversion process is broken down into a succession of steps. It starts with the efficiencies of heat generation and transfer in the heat generation part of the schematic shown in Figure 4.1. In a second step the efficiency of the heat engine and the cooling part are discussed. Both parts are again multistage processes where the initial heat generation part involves the processes combustion, heat exchange, and stack gas release, and in the following heat engine part the four steps of the Rankine cycle as described in Section 4.1.1. An example calculation using realistic values will then be given further down in Section 4.1.3.

Combustion: The combustion process converts the chemical energy stored in the fuels into hot flue gas. The first law efficiency of the combustion process is $\eta_{1,\text{comb}} = 1$, as the energy stored in the chemical bonds is defined as the heating value of these fuels, as described in Section 1.1. The second-law efficiency of the combustion process will be discussed in Chapter 6. For the moment a typical value of $\eta_{2,\text{comb}} \approx 0.7$ is assumed. The reason for the exergy losses is the irreversibility of the combustion process. In a solar thermal power plant, the efficiencies in this step correspond to the solar energy collection first- and second-law efficiencies which will be discussed in detail in Chapter 9. The efficiency considerations in the rest of this section can then also be used for this type of power plant.

Heat exchange: The hot flue gas passes the steam generator, where it transfers an amount of heat ${}_2q_3$ per unit mass working fluid into the working fluid of the Rankine cycle in a closed heat exchanger. Assuming the heat exchanger to be ideally insulated against the surroundings, again no energy is lost, and $\eta_{1,\text{sg}} = 1$ holds, as all energy leaving the hot flue gas in the heat exchanger is also entering the working fluid. Note that only the heat-exchange process is considered in this step; the losses due to the elevated temperature of the cooled flue gas leaving the heat exchanger are subject of the next point. The second-law efficiency of the heat exchange process is given by the ratio of the exergy taken up by the working fluid in the isobaric heat exchange $\psi(\,{}_2q_3)$ and the exergy loss of the flue gas when it streams through the heat exchanger:

$$\eta_{2,\mathrm{sg}} = \frac{\psi({}_2q_3)}{\psi_{\mathrm{in,fg}} - \psi_{\mathrm{stack}}}. \tag{4.10}$$

Here $\psi_{\mathrm{in,fg}}$ denotes the exergy of the flue gas entering the heat exchanger, and ψ_{stack} is its exergy when it leaves the heat exchanger to be released as stack gas. Both quantities are specific to the mass flow of the working fluid, not of the flue gas. $\eta_{2,\mathrm{sg}}$ can be calculated using equation (3.19) or equation (3.26) and is typically of the order of $\eta_{2,\mathrm{sg}} \approx 0.7$. The heat exchange process is one of the main exergy loss mechanisms of the steam power plant, the high losses being due to the typically large temperature difference between the hot flue gas and the working medium.

Stack gas release: After passing the heat exchanger, the flue gas will in general have retained a certain temperature $T = T_{\mathrm{stack}}$ above the ambient temperature. Parts of the input heat $q_{\mathrm{in,fg}}$ and exergy $\psi_{\mathrm{in,fg}}$ from the combustion process are thereby discarded in the form of heat q_{stack} released through the stack to the atmosphere. The first- and second-law efficiencies of this process are then given by the ratios of the net energy and exergy transfers to the working fluid and the input energy and exergy of the hot flue gas, respectively:

$$\begin{aligned}\eta_{1,\mathrm{stack}} &= \frac{q_{\mathrm{in,fg}} - q_{\mathrm{stack}}}{q_{\mathrm{in,fg}}} = \frac{{}_2q_3}{q_{\mathrm{in,fg}}}\\ \eta_{2,\mathrm{stack}} &= \frac{\psi_{\mathrm{in,fg}} - \psi_{\mathrm{stack}}}{\psi_{\mathrm{in,fg}}} = 1 - \frac{\psi_{\mathrm{stack}}}{\psi_{\mathrm{in,fg}}} = 1 - \frac{h_{\mathrm{stack}} - h_{\mathrm{am}} - T_{\mathrm{am}}(s_{\mathrm{stack}} - s_{\mathrm{am}})}{h_{\mathrm{in,fg}} - h_{\mathrm{am}} - T_{\mathrm{am}}(s_{\mathrm{in,fg}} - s_{\mathrm{am}})}.\end{aligned} \tag{4.11}$$

The three steps mentioned so far constitute the full first- and second-law efficiencies of the heat generation process, i.e. of the conversion of energy stored in chemical fuels to the input heat ${}_2q_3$ of the Rankine cycle.

In a next step the total conversion efficiency of the heat engine part is discussed, assuming an ideal Rankine cycle. For an ideal reversible process with $T_{4,1} = T_{\mathrm{am}}$, the first-law efficiency of the entire cycle is given by equation (4.8), and the second-law efficiency is $\eta_{2,\mathrm{id.Rank}} = 1$. If such an ideal reversible heat engine is assumed, the overall first- and second-law efficiencies of the entire power plant, i.e. the first- and second-law efficiencies for the conversion of chemical energy into technical work, can be calculated:

$$\begin{aligned}\eta_1 &= \underbrace{\eta_{1,\mathrm{comb}}}_{=1} \cdot \underbrace{\eta_{1,\mathrm{sg}}}_{=1} \cdot \eta_{1,\mathrm{stack}} \cdot \left(1 - \frac{T_{\mathrm{am}}}{\overline{T}_{\mathrm{in}}}\right) = \eta_{1,\mathrm{stack}} \cdot \left(1 - \frac{T_{\mathrm{am}}}{\overline{T}_{\mathrm{in}}}\right)\\ \eta_2 &= \eta_{2,\mathrm{comb}} \cdot \eta_{2,\mathrm{sg}} \cdot \eta_{2,\mathrm{stack}} \cdot \underbrace{\eta_{2,\mathrm{id.Rank}}}_{=1} = \eta_{2,\mathrm{comb}} \cdot \eta_{2,\mathrm{sg}} \cdot \eta_{2,\mathrm{stack}}.\end{aligned} \tag{4.12}$$

If the Rankine cycle is not ideal and reversible, as is the case for any real machine, further energy and exergy loss mechanisms have to be considered. The output of electrical energy from the power plant is reduced by the mechanical efficiency of the turbine and the generator, i.e. the ratio between the energy loss of the working fluid in the turbine and the electrical energy produced. For this ratio a value of $\eta_{1,turb} = \eta_{2,turb} = \eta_{turb}$ typically in the order of $\eta_{turb} \approx 0.9$ is realistic. The first- and second-law efficiencies for this process are identical, as only mechanical and electrical energy, i.e. pure exergy, are involved in the energy conversion process. The feed water pump has a similar mechanical efficiency, which can, however, be neglected, as only a small fraction of the energy provided by the turbine is required to drive the feedwater pump. Furthermore, heat losses from the flue gas to the surroundings are still assumed to be negligible. Then, for the overall first-law efficiency of the conversion of fuels into electricity in a steam power plant, the following expression holds:

$$\begin{aligned}\eta_1 = \frac{w_{out,mech}}{\Delta h_{r;in,fuel}} &= \eta_{1,comb} \cdot \frac{w_{out,mech}}{q_{in,fg}} = \eta_{1,comb} \cdot \eta_{1,sg} \cdot \eta_{1,stack} \cdot \frac{w_{out,mech}}{{}_2q_3} \\ &= \underbrace{\eta_{1,comb}}_{=1} \cdot \underbrace{\eta_{1,sg}}_{=1} \cdot \eta_{1,stack} \cdot \eta_{mech} \cdot \left(1 - \frac{T_{4,1}}{\overline{T}_{in}}\right) \\ &= \eta_{1,stack} \cdot \eta_{mech} \cdot \left(1 - \frac{T_{4,1}}{\overline{T}_{in}}\right),\end{aligned} \tag{4.13}$$

where $\Delta h_{r;in,fuel}$ is the heating value of the input fuel per unit mass working fluid, and $q_{in,fg}$ is the heat per unit mass working fluid transferred into the flue gas during combustion. Energy losses thus only occur due to the elevated temperature of the gas released through the stack, mechanical losses, and the fact that that the entropy entering the Rankine cycle with the input heat has to be rejected at the lower temperature level $T_{4,1}$. By far the most important contribution to the losses is the latter, i.e. the fact that even an ideal heat engine following a Rankine cycle can only convert a part of the input heat energy to mechanical energy. The energy efficiency is hence limited by the fact that a heat engine is used, and thus by the efficiency restrictions the second law of thermodynamics dictates for such engines. The stack losses, and also a part of the cooling losses, can be attenuated by combining the power plant with a remote heating device, where the low exergy heat rejected in both loss mechanisms is used for remote room heating.

If the Rankine cycle is run at a low temperature level of $T_{4,1} = T_{am}$, there is no exergy loss in the condenser, as the rejected heat is then pure anergy. For $T_{4,1} > T_{am}$ an amount of exergy $\psi_{loss,cond}$ is lost in the isothermal cooling process:

$$\psi_{loss,cond} = {}_4q_1 \left(1 - \frac{T_{am}}{T_{4,1}}\right) = \Delta s(T_{4,1} - T_{am}), \tag{4.14}$$

where $\Delta s = s_3 - s_2 = s_4 - s_1$ is the entropy transported by the input and output heats ${}_2q_3$ and ${}_4q_1$, respectively. The effect of the cooling process on the overall second-law efficiency of the Rankine cycle is then the ratio of the actual net input exergy at the chosen parameters and the full input exergy in the isobaric heating step 2 → 3, which is by far smaller than the exergy loss in the steam generator:

$$\eta_{2,\mathrm{cond}} = \frac{\psi({}_2q_3) - \psi_{\mathrm{loss,cond}}}{\psi({}_2q_3)} = \frac{\Delta s(\overline{T}_{\mathrm{in}} - T_{\mathrm{am}} - T_{4,1} + T_{\mathrm{am}})}{\Delta s(\overline{T}_{\mathrm{in}} - T_{\mathrm{am}})} = \frac{\overline{T}_{\mathrm{in}} - T_{4,1}}{\overline{T}_{\mathrm{in}} - T_{\mathrm{am}}}. \tag{4.15}$$

The overall second-law efficiency of a steam power plant is correspondingly found to be

$$\begin{aligned}\eta_2 = \frac{w_{\mathrm{out,mech}}}{\psi_{\mathrm{in,fuel}}} &= \eta_{2,\mathrm{comb}} \cdot \frac{w_{\mathrm{out,mech}}}{\psi_{\mathrm{in,fg}}} = \eta_{2,\mathrm{comb}} \cdot \eta_{2,\mathrm{sg}} \cdot \eta_{2,\mathrm{stack}} \frac{w_{\mathrm{out,mech}}}{{}_2\psi_3}\\ &= \eta_{2,\mathrm{comb}} \cdot \eta_{2,\mathrm{sg}} \cdot \eta_{2,\mathrm{stack}} \cdot \eta_{\mathrm{mech}} \frac{-({}_1w_2 + {}_3w_4)}{\psi({}_2q_3)}\\ &= \eta_{2,\mathrm{comb}} \cdot \eta_{2,\mathrm{sg}} \cdot \eta_{2,\mathrm{stack}} \cdot \eta_{2,\mathrm{cond}} \cdot \eta_{\mathrm{mech}} \underbrace{\frac{-({}_1w_2 + {}_3w_4)}{\psi({}_2q_3) - \psi_{\mathrm{loss,cond}}}}_{=1}\\ &= \eta_{2,\mathrm{comb}} \cdot \eta_{2,\mathrm{sg}} \cdot \eta_{2,\mathrm{stack}} \cdot \eta_{2,\mathrm{cond}} \cdot \eta_{\mathrm{mech}},\end{aligned} \tag{4.16}$$

where $\psi_{\mathrm{in,fuel}}$ is the exergy transferred into the system by the fuel required to provide the heat ${}_2q_3$ per unit mass of working fluid and $\psi_{\mathrm{in,fg}}$ the exergy of the flue gas produced by the burning of the fuel. The last factor in the third line of equation (4.16) is equal to 1, as it is the ratio between net input and output exergies of an ideal reversible Rankine cycle. All energy- and exergy-loss mechanisms discussed above are schematically shown in an exergy flow diagram in Figure 4.5.

The energy efficiency of the individual energy conversion steps can be read out by comparing the input and output full widths of the arrows, and the exergy efficiencies by only comparing the white exergy parts of the arrows. It can be readily seen that the

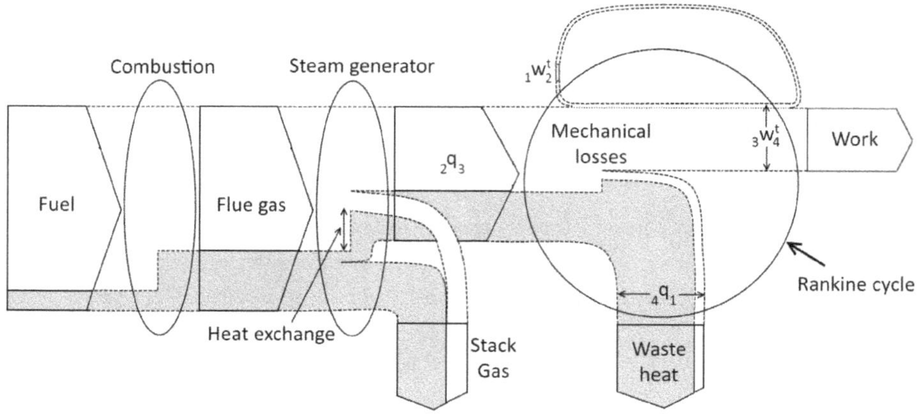

Fig. 4.5: Energy and exergy flow diagram for the energy conversion processes in a steam power plant. The *ellipsoids* represent energy conversion processes and the *arrows* represent transferred energy, with the upper white fraction of the *arrows* representing the transferred exergy.

main energy losses in the steam power plant occur due to the heat rejections in the cooling step of the Rankine cycle and the stack gas release. Combustion and the heat exchange in the steam generator do not lead to energy losses. A contrary picture arises when the exergy loss mechanisms are considered. Here the main losses occur during the combustion process and, most importantly, during the heat exchange in the steam generator. In the latter process the substantial exergy loss is caused by the relatively large temperature gradient across the heat exchanger which is required to generate the desired amount of power.

4.1.3 Example calculation

As an example, the first- and second-law efficiencies of a coal-fired power plant with realistic parameters are calculated. The following working parameters are assumed: $T_{max} = 650\,°C \Rightarrow \overline{T}_{in} \approx 350\,°C = 623$ K; $T_{4,1} = 35\,°C = 308$ K; $T_{in,fg} = 1200\,°C = 1473$ K; $T_{stack} = 200\,°C = 473$ K; and $p_3 = 275$ bar $= 27.5$ MPa. Note that for the parameters chosen, the steam generator is operated in the supercritical region. This is only of minor importance, and the calculations still give a good impression of the steam power plant efficiencies.

Heat exchange in the steam generator: Feed water heating and steam generation are both achieved with heat exchangers which together increase the feed water temperature from $T_2 \approx T_{am} = 25\,°C = 298$ K to $T_3 = 650\,°C = 923$ K at a constant pressure of $p = 275$ bar $= 27.5$ MPa. For the average temperature of the working fluid, $\overline{T}_{in} = 350\,°C = 623$ K is assumed. The values for the mass specific enthalpies and entropies of these points are $h_2 = 170$ J/g and $s_2 = 0.50$ J/gK at the steam generator inlet and $h_3 = 3600$ J/g and $s_3 = 6.5$ J/gK at the steam generator outlet (source: David R. Lide, ed., *CRC Handbook of Chemistry and Physics, Internet Version 2005*, http://www.hbcpnetbase.com, CRC Press, Boca Raton, FL, 2005). According to equation (3.16) the total exergy input in the steam generator is then given by

$$ {}_2\psi_3 = \psi_3^x - \psi_2^x = (h_3 - h_2) - T_{am}(s_3 - s_2) = 1600\ \mathrm{J/g}. \tag{4.17} $$

At the same time the flue gas cools down from an estimated temperature of $T_{in,fg} = 1200\,°C = 1473$ K to $T_{stack} = 200\,°C = 473$ K at atmospheric pressure. The flue gas is treated as an ideal gas with five motional degrees of freedom leading to $c_p = 7/2R = 1.01$ J/gK, which is a reasonable assumption for the temperature range considered (for details on the heat capacity of ideal gases see Section A.2 of the appendix). For the calculation of the second-law efficiency of the heat exchangers, values specific to the same mass flows have to be chosen. The working fluid water is used as this reference mass flow. The overall energy leaving the flue gas per unit time $\dot{Q}$ is identical to the overall energy entering the working fluid water per unit time, as the heat exchanger has a first law efficiency of $\eta_{1,\text{sg}} = 1$. Therefore, the respective mass flows can be adjusted using:

$$\begin{aligned} \dot{Q} &= \dot{m}_{\text{gas}} c_{\text{p}} (T_{\text{in,fg}} - T_{\text{stack}}) = \dot{m}_{\text{water}} (h_3 - h_2) \\ \Rightarrow \frac{\dot{m}_{\text{gas}}}{\dot{m}_{\text{water}}} &= \frac{(h_3 - h_2)}{c_{\text{p}} (T_{\text{in,fg}} - T_{\text{stack}})} . \end{aligned} \tag{4.18}$$

The exergy per unit mass water working fluid leaving the flue gas in the heat exchanger can now be calculated via equations (3.20) and (3.21):

$$\begin{aligned} \psi_{\text{in,fg}} - \psi_{\text{stack}} &= \frac{\dot{m}_{\text{gas}}}{\dot{m}_{\text{water}}} \cdot \int_{T_{\text{stack}}}^{T_{\text{in,fg}}} c_{\text{p}} \left(1 - \frac{T_{\text{am}}}{T}\right) dT \\ &= \frac{(h_3 - h_2)}{(T_{\text{in,fg}} - T_{\text{stack}})} \left(T_{\text{in,fg}} - T_{\text{stack}} - T_{\text{am}} \ln \left(\frac{T_{\text{in,fg}}}{T_{\text{stack}}} \right) \right) \\ &= 2200\,\text{J/g} \end{aligned} \tag{4.19}$$

The overall second-law efficiency of the heat exchange during the feed water heating and in the steam generator is then given by

$$\eta_{2,\text{sg}} = \frac{{}_2\psi_3}{\psi_{\text{in,fg}} - \psi_{\text{stack}}} = 0.72 \tag{4.20}$$

Stack gas release: The first- and second-law efficiencies associated with the energy and exergy losses through hot stack gas can be calculated using equation (4.11):

$$\begin{aligned} \eta_{1,\text{stack}} &= \frac{c_{\text{p}} (T_{\text{in,fg}} - T_{\text{stack}})}{c_{\text{p}} (T_{\text{in,fg}} - T_{\text{am}})} = 1 - \frac{T_{\text{stack}} - T_{\text{am}}}{T_{\text{in,fg}} - T_{\text{am}}} = 0.85 \\ \eta_{2,\text{stack}} &= 1 - \frac{\psi_{\text{stack}}}{\psi_{\text{in,fg}}} = 1 - \frac{T_{\text{stack}} - T_{\text{am}} - T_{\text{am}} \ln (T_{\text{stack}}/T_{\text{am}})}{T_{\text{in,fg}} - T_{\text{am}} - T_{\text{am}} \ln (T_{\text{in,fg}}/T_{\text{am}})} = 0.95. \end{aligned} \tag{4.21}$$

The stack gas thus carries 15 % of the energy of the flue gas out of the plant but only 5 % of its exergy.

Cooling process: The exergy loss per unit mass working fluid in the cooling process $\psi_{\text{loss,cond}}$ can be calculated, assuming a temperature of $T_{4,1} = 35\,°\text{C}$ in the condenser. According to equation (4.15), this leads to the following second-law efficiency of the condenser:

$$\eta_{2,\text{cond}} = \frac{\overline{T}_{\text{in}} - T_{4,1}}{\overline{T}_{\text{in}} - T_{\text{am}}} = 0.97. \tag{4.22}$$

With the values calculated above, the overall first- and second-law efficiencies can be obtained using equations (4.13) and (4.16). Therefore the realistic values of $\eta_{2,\text{comb}} = 0.7$

and $\eta_{mech} = 0.9$ are assumed. The overall first- and second-law efficiencies then become:

$$\eta_1 = 0.85 \cdot 0.9 \cdot \left(1 - \frac{308}{623}\right) = 0.41$$
$$\eta_2 = 0.7 \cdot 0.72 \cdot 0.95 \cdot 0.9 \cdot 0.97 = 0.42. \tag{4.23}$$

If the combustion process is neglected and the hot flue gas is taken as the initial energy input, the second-law efficiency increases to $\eta_2 = 0.60$. It is then clearly dominated by the irreversibilities in the heat exchanger of the steam generator.

4.1.4 Modifications in a real steam power plant

A real power plant does not simply operate along a Rankine cycle, but along a modified version of the cycle to improve the technical feasibility and to increase the efficiency. The modifications are mostly necessitated by the technical limitations of the devices and materials used. Next to these necessary conditions, they are aimed at an increased average input temperature $\overline{T}_{in}$ and thus a higher overall first-law efficiency. Modern steam power plants are often working in the supercritical region where the working fluid does not undergo a phase transition as the liquid and the gaseous phase are no longer well discernible. This fact, however, changes nothing in the principle design of the power plants and the modifications mentioned in this section.

Superheat: As discussed above (see equation (4.7)), the higher $\overline{T}_{in}$, the higher is the first-law efficiency. In an unmodified Rankine cycle a higher $\overline{T}_{in}$ implies a higher upper pressure level and thus a higher boiling temperature. However, from a practical point of view, a higher boiling temperature also leads to a difficulty for the turbine operation. As can be seen below in Figure 4.6, the steam mass fraction x_4 at the turbine outlet decreases with an increasing boiling point, and there is a lower limit to the steam mass fraction, below which the turbine erodes quickly. To meet both demands, the high desired value of $\overline{T}_{in}$ and the relatively high steam mass fraction x_4, the steam at point 3 has to be heated further before it enters the turbine inlet. This process is called **superheating**, and its effect on the thermodynamic cycle is shown in Figure 4.6.
The superheating has the additional effect that the average heating temperature $\overline{T}_{in}$ increases even further. Note that the steam mass fraction x_4 does not have to be equal to 1, because turbines can handle moderate amounts of liquid water.

Reheat: The materials used for the tubing and the turbine itself only tolerate temperatures up to a characteristic $T = T_{max}$. The optimization criterion in the superheat process was the steam mass fraction at the turbine outlet x_4. With an optimal value for x_4, a lowering of the pressure p_3 might then be necessary to keep the temperature $T_3 \leq T_{max}$ without further modifications. This can be omitted by a two-stage turbine in combination with a step called **reheating**. In this process

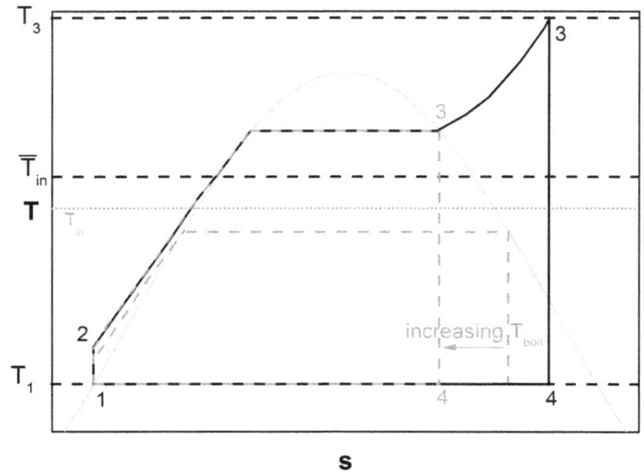

Fig. 4.6: Ts-diagram of a thermodynamic cycle describing a Rankine cycle modified with superheating (*thick solid line*) and two unmodified Rankine cycles (*dashed line*). The average input temperatures of the modified Rankine cycle and the unmodified Rankine cycle with the same upper pressure level are also shown.

the superheated steam is first relaxed through a high pressure turbine from the highest pressure level p_3 to an intermediate pressure level p_{3a} and then heated again. After this reheating it is relaxed to the lowest pressure level p_4 in a second low-pressure turbine. An additional advantage of the reheating process is that the pressure level p_{3a} offers another free parameter which can be used to optimize the power plant further. In a real power plant typically at least one reheating step is implemented, as shown in Figure 4.7.

Note that the reheating process leads to a more isothermal heating process, because the multistep isobaric heating curves 2 → 3 and 3a → 3b in Figure 4.7 do not deviate as much from an average temperature as, for example, 2 → 3 in Figure 4.6. A more isothermal heat transfer is beneficial in terms of efficiency, as the

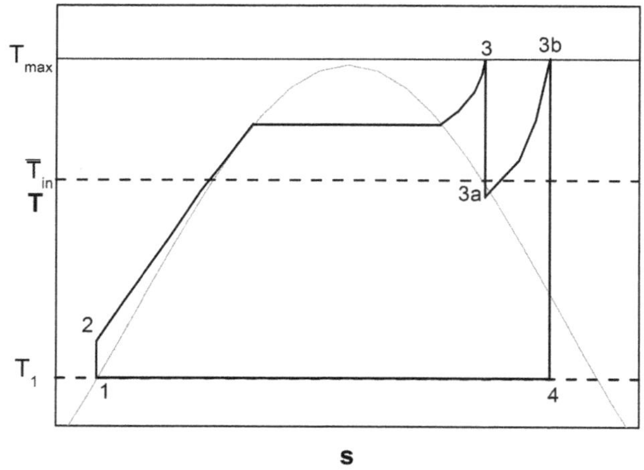

Fig. 4.7: Ts-diagram of a thermodynamic cycle describing a Rankine cycle modified with superheating and reheating.

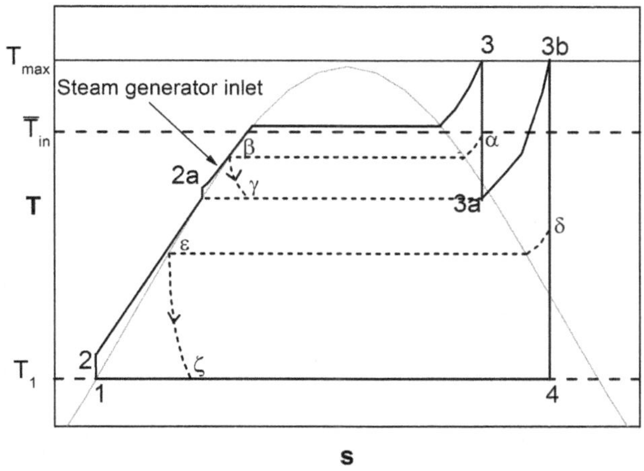

Fig. 4.8: Ts-diagram of a thermodynamic cycle describing a steam power plant with superheating, reheating and a regenerative feed water heating (*solid line*: feed water line; *dashed line*: diverted water for regenerative feed water heating).

Rankine cycle then approaches the Carnot cycle. A succession of many reheating steps is thus called **carnotization.**

Regenerative feed water heating: The major reason for the relatively low second-law efficiency of the steam generator is the high temperature difference between the combustion gases and the working fluid of the steam power plant, as discussed in Section 4.1.2. To mitigate this problem, part of the water vapor is taken out of the main feed water line at several points in the cycle to preheat the liquid water in heat exchangers. The feed water then reaches the inlet of the steam generator at a temperature already close to the boiling point. This modification increases $\overline{T}_{in}$, and the correspondingly modified Rankine cycle is shown in Figure 4.8.

For diverting part of the vapor stream, the two turbines are segmented into several stages each (in Figure 4.8 two stages are shown), and additional pressure levels $p_{3a} < p_{\alpha,\beta} < p_3$ and $p_4 < p_{\delta,\varepsilon} < p_{3a}$ are thus introduced. The heating of the liquid water can be achieved either by closed or open feed water heaters. Closed feed water heaters are counterflow heat exchangers where the part of the steam taken out of the main line cools down and condenses. In Figure 4.8 this corresponds to the top and the bottom *dashed lines* at pressure levels $p_{\alpha,\beta}$ and $p_{\delta,\varepsilon}$, respectively. In an open feed water heater the relatively hot diverted vapor stream and the relatively cold feed water are mixed. This has the advantage that no temperature gradient has to be maintained, as is the case for a closed feed water heater. To mix two material flows it is necessary for them to have the same pressure. For this reason the feed water pump is divided into two stages, where the first stage increases the pressure from p_1 to $p_{2,3a}$ before the open feed water heater, and the second stage increases the feed water pressure from $p_{2,3a}$ to $p_{2a,3}$. A sketch of a thus modified steam power plant is shown in Figure 4.9

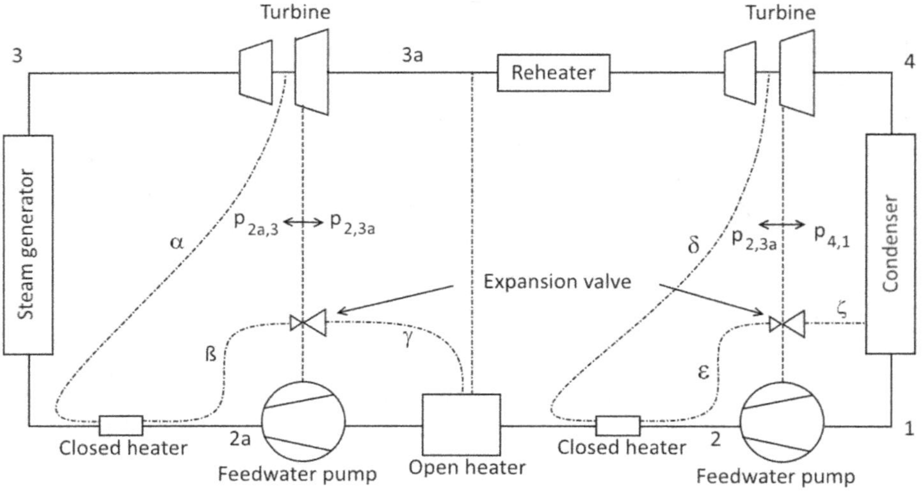

Fig. 4.9: Schematic of a steam power plant with all modifications described in this section. The main feed water line is represented by the *solid line* while the diverted vapor lines are represented by *dash-dotted lines*. The *vertical dashed lines* separate the three main pressure levels in the cycle.

4.2 Gas turbine power plants

Most gas turbine power plants are open gas turbines. They are designed in a very different, more compact way than the steam power plant. A schematic is shown in Figure 4.10.

There are not three different subsystems for heating, conversion from heat to electricity in a heat engine, and cooling as in the steam power plant. Instead, all essential components for the heat engine and the conversion of heat into electrical energy, i.e. compressor, heating chamber, turbine, and generator, can be mounted on a single shaft. The working fluid is air and does not change its phase, but is in the gaseous phase throughout the whole thermodynamic cycle. It does, however, change its composition by the addition and subsequent burning of the gaseous fuel. The fuel used is typically natural gas, i.e. methane, but also other gases, for example hydrogen, can be used. The high hydrogen content of the methane leads to a high water content in the products of the combustion reaction and, consequently, a relatively low share of CO_2. The details of the chemical reaction underlying the combustion process will be discussed in Chapter 6. In an open cycle gas turbine power plant, heat exchangers do not have to be used. In the combustion chamber the natural gas is added to the working fluid and burnt using a part of the oxygen content of the working fluid air. Since the hot flue gases directly expand through the turbine, the exergy losses connected with the transfer of heat onto the working fluid, which amount to about half

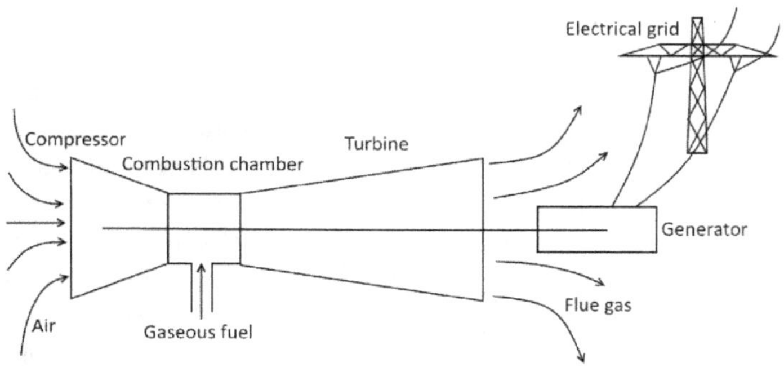

Fig. 4.10: Schematic of a gas turbine power plant.

of the overall exergy losses of a steam power plant (see equation (4.23)), do not occur in the gas turbine power plant. Also the second heat exchanger, which is used for the cooling, is realized rather elegantly by the continuous exchange of matter with the, approximately infinite, surrounding heat bath. However, as will be shown below, the high temperature of the gas at the outlet of the turbine leads to substantial losses. The appropriate, open operation cycle is shown schematically in Figure 4.11.

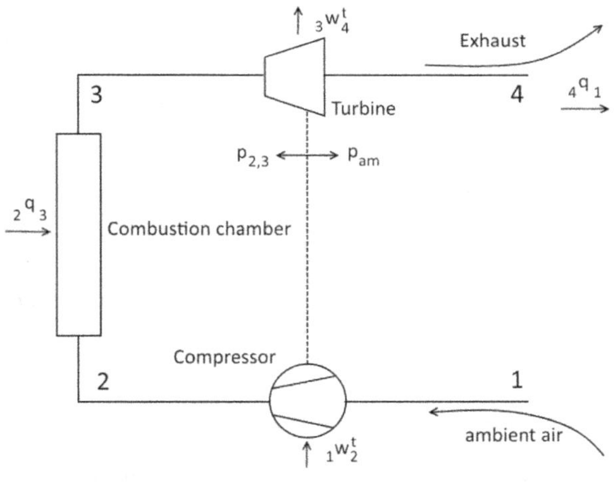

Fig. 4.11: Schematic of a gas turbine power plant in an open, realistic configuration. The *vertical dashed line* separates the two pressure levels in the cycle.

An important implication of the open process realization for calculations is that state 1 at the compressor inlet is always the ambient state, and the lower pressure level therefore always $p_{4,1} = p_{am}$.

4.2.1 Joule–Brayton cycle

In the following, the components of a gas turbine power plant and the thermodynamic processes will be described under two assumptions which are typically not realized in real power plants, but make the description of the cycle and comparison to the Rankine cycle easier. First, it is assumed that the working fluid is air throughout the cycle, i.e. its change of composition due to mixing with the gaseous fuel and the combustion process is neglected; and, second, that the heating of the air is taking place in a closed heat exchanger. Furthermore, since the gas discharged to the atmosphere at the turbine outlet is eventually brought to the same state as the ambient air which enters the compressor inlet, this process can be modeled, assuming that the exhaust gas is brought to ambient conditions by passing through a closed heat exchanger. In this configuration, shown in Figure 4.12, the process is closed as the same amount of air is cycling through each of the four components.

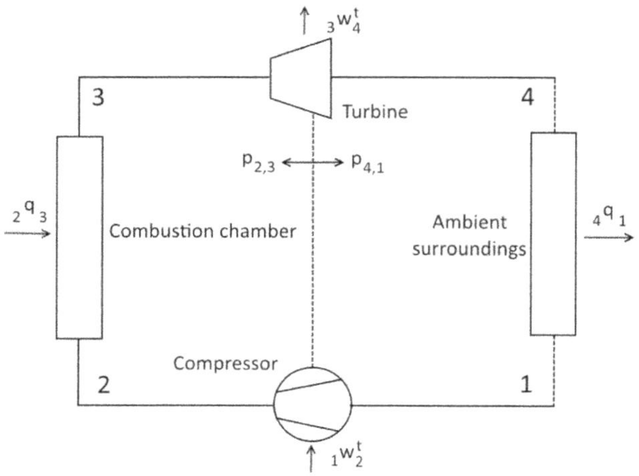

Fig. 4.12: Schematic of a gas turbine power plant in a closed configuration. The *vertical dashed line* separates the two pressure levels in the cycle.

The individual state changes of the working fluid are the following:

1 → 2: Air at ambient conditions is fed to a **compressor**, where it is compressed to the higher pressure level of $p = p_{2,3}$. This process requires a significant amount of technical work $_1w_2^t$, as the working fluid air has a relatively low volume density, i.e. a relatively high mass-specific volume. The technical work required for the compression can be calculated using equation (3.9):

$$_1w_2^t = h_2 - h_1 = \int_{p_{4,1}}^{p_{2,3}} v(p)dp = \int_{p_{4,1}}^{p_{2,3}} 1/\varrho(p)dp > 0. \qquad (4.24)$$

2 ⟶ 3: The compressed air takes up the heat ${}_2q_3$ from the combustion gases in the **combustion chamber**, which is modeled as a closed heat exchanger as mentioned above. The process is isobaric, which means that no technical work is exchanged during the process:

$$ {}_2q_3 = h_3 - h_2 = \int_{T_2}^{T_3} c_\mathrm{p}(T)dT > 0. \tag{4.25} $$

3 ⟶ 4: The heated gas is adiabatically expanded in the **turbine** from the higher pressure level $p = p_{2,3}$ to the lower pressure level $p = p_{4,1}$. During this process the turbine performs the specific technical work ${}_3w_4^\mathrm{t}$ at the electrical generator. According to the first law of thermodynamics for open systems in a steady flow equilibrium, equation (3.9), this leads to

$$ {}_3w_4^\mathrm{t} = h_4 - h_3 = \int_{p_{2,3}}^{p_{4,1}} 1/\varrho(p)dp < 0. \tag{4.26} $$

4 ⟶ 1: The relaxed gas is fed to the heat exchanger, where it is cooled down isobarically to ambient temperature releasing the waste heat ${}_4q_1$ to the surroundings. The final state 1 of this process is the ambient state:

$$ {}_4q_1 = h_1 - h_4 = \int_{T_4}^{T_1} c_\mathrm{p}(T)dT < 0. \tag{4.27} $$

The closed configuration of the ideal gas turbine power plant is a realization of the **Joule–Brayton cycle**, the Ts- and pv-diagrams of which are shown in Figure 4.13.

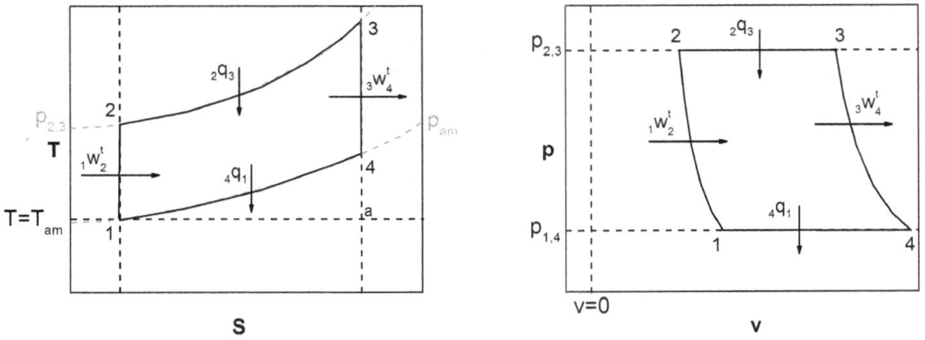

Fig. 4.13: Ts- and pv-diagram of the ideal Joule–Brayton cycle consisting of two isobars realized in the surrounding atmosphere (4 ⟶ 1) and the heating chamber (2 ⟶ 3) linked by two adiabatic state changes realized in the compressor (1 ⟶ 2) and the turbine (3 ⟶ 4).

The first-law efficiency of an ideal gas turbine power plant following the Joule–Brayton cycle can be calculated analogously to the one of steam power plants, as given in equation (4.5):

$$\eta_1 = \frac{-({}_1w_2^t + {}_3w_4^t)}{{}_2q_3} = \frac{{}_2q_3 + {}_4q_1}{{}_2q_3} = 1 - \frac{|{}_4q_1|}{|{}_2q_3|} = 1 - \frac{\int_{T_1}^{T_4} c_p(T)dT}{\int_{T_2}^{T_3} c_p(T)dT}. \tag{4.28}$$

To get an estimate of the value for the first-law efficiency η_1, a constant c_p for all temperatures involved is assumed. This assumption offers a good approximation, as c_p is only changing slightly in the temperature region considered. This leads to

$$\eta_1 = 1 - \frac{c_p(T_4 - T_1)}{c_p(T_3 - T_2)} = 1 - \frac{T_1}{T_2} \cdot \frac{(T_4/T_1 - 1)}{(T_3/T_2 - 1)}. \tag{4.29}$$

As the processes $3 \to 4$ and $1 \to 2$ are both adiabatic processes linking two isobars, the following expression holds (see equation (A.20) in the appendix):

$$\frac{T_4}{T_3} = \left(\frac{p_{4,1}}{p_{2,3}}\right)^{1-1/\gamma} = \frac{T_1}{T_2} \tag{4.30}$$

$$\Rightarrow \frac{T_4}{T_1} = \frac{T_3}{T_2}$$

$$\Rightarrow \eta_1 = 1 - \frac{T_1}{T_2} = 1 - \frac{T_4}{T_3} = 1 - \left(\frac{p_{4,1}}{p_{2,3}}\right)^{1-1/\gamma} = 1 - \left(\frac{p_{2,3}}{p_{4,1}}\right)^{1/\gamma-1} =: 1 - \pi^{1/\gamma-1},$$

with the adiabatic exponent $\gamma := c_p/c_v$. In this formulation it becomes evident that the first-law efficiency for the idealized gas turbine power plant only depends on the ratio π between the upper pressure level $p = p_{2,3}$ and the lower pressure level $p = p_{4,1}$. Contrary to the Rankine cycle, the cooling process is not realizable as an isothermal process at $T = T_{am}$, as no phase transition of the working fluid is occurring. A fraction of the exergy input is thus necessarily discarded in the cooling process, as can be seen by comparing the area under the curve $4 \to 1$ in the Ts-diagram in Figure 4.13, with the corresponding rectangle below the *dashed line* at $T = T_{am}$. The former is identical to the heat ${}_4q_1$ rejected, while the latter is its anergy content $A({}_4q_1) = T_{am}(s_4 - s_1)$. The discarded exergy is thus given by the area $\overline{14a1}$. For the maximal possible second-law efficiency the net exergy input is important, i.e. the exergy transported into the system by ${}_2q_3$ minus the exergy discarded in ${}_4q_1$. This necessary exergy output in the cooling process is a major drawback of the gas turbine power plant compared to the steam power plant.

4.2.2 Optimization criteria

As for any real gas turbine power plant, the maximal temperature T_3 is limited by the material properties of the gas turbine and the tubing, and the minimal temperature is always the ambient temperature $T_1 = T_{am}$, the pressure ratio $\pi = p_{2,3}/p_{4,1} =$

$p_{2,3}/p_{\mathrm{am}}$ becoming the only degree of freedom for the process description. An important quantity for the process optimization is the ratio between turbine power output and compressor power consumption, the so-called **back-work ratio**, which can be calculated using equations (A.19) and (A.20) from the appendix for adiabatic state changes:

$$
\begin{aligned}
{}_1w_2^{\mathrm{t}} &= \int_{p_{\mathrm{am}}}^{p_{2,3}} 1/\varrho(p)dp \\
&= \frac{(p_{\mathrm{am}})^{1/\gamma}}{\varrho_{\mathrm{am}}} \int_{p_{\mathrm{am}}}^{p_{2,3}} p^{-1/\gamma}dp = \frac{(p_{\mathrm{am}})^{1/\gamma}}{\varrho_{\mathrm{am}}} \frac{\gamma}{\gamma-1}((p_{2,3})^{1-1/\gamma} - (p_{\mathrm{am}})^{1-1/\gamma}) \\
|{}_3w_4^{\mathrm{t}}| &= \int_{p_{\mathrm{am}}}^{p_{2,3}} 1/\varrho(p)dp \\
&= \frac{(p_{2,3})^{1/\gamma}}{\varrho_3} \int_{p_{\mathrm{am}}}^{p_{2,3}} p^{-1/\gamma}dp = \frac{(p_{2,3})^{1/\gamma}}{\varrho_3} \frac{\gamma}{\gamma-1}((p_{2,3})^{1-1/\gamma} - (p_{\mathrm{am}})^{1-1/\gamma}) \\
\Rightarrow \frac{|{}_1w_2^{\mathrm{t}}|}{|{}_3w_4^{\mathrm{t}}|} &= \frac{\varrho_3}{\varrho_{\mathrm{am}}}\pi^{-1/\gamma} = \left(\frac{p_{2,3}T_1}{p_{\mathrm{am}}T_3}\right)\pi^{-1/\gamma} = \pi^{1-1/\gamma}\frac{T_1}{T_3}.
\end{aligned}
\tag{4.31}
$$

This means that an increased pressure ratio leads to an increased fraction of the turbine output energy being necessary to run the compressor. At a limiting pressure ratio of $\pi = (T_3/T_1)^{\gamma/(\gamma-1)}$ the absolute values of the turbine and compressor work become identical, and $T_2 = T_3$ holds. At this point no net work output can be obtained, i.e. $w_{\mathrm{out}}^{\mathrm{t}} = 0$. The same is true for the other limiting pressure ratio $\pi = 1$, as the process is then no longer working between two different pressure levels, and the net heat input becomes zero. Between the two limiting values $1 < \pi < (T_3/T_1)^{\gamma/(\gamma-1)}$ with $w_{\mathrm{out}}^{\mathrm{t}} = 0$, a net work $w_{\mathrm{out}}^{\mathrm{t}} > 0$ can be extracted, and a maximum exists. This net work output optimized pressure ratio $\pi_{\mathrm{max,w}}$ can then be found for a given temperature ratio $T_3/T_1 =: \tau$ by writing the work output as a function of the pressure ratio π and the temperature ratio τ:

$$
\begin{aligned}
w_{\mathrm{out}}^{\mathrm{t}} &= {}_1w_2^{\mathrm{t}} + {}_3w_4^{\mathrm{t}} = -({}_2q_3 + {}_4q_1) = -c_{\mathrm{p}}\left((T_3 - T_2) + (T_1 - T_4)\right) \\
&= -c_{\mathrm{p}}T_1\left(\tau - \pi^{1-1/\gamma} + 1 - \frac{\tau}{\pi^{1-1/\gamma}}\right).
\end{aligned}
\tag{4.32}
$$

Substituting $y := \pi^{1-1/\gamma}$ and evaluating $(\partial w_{\mathrm{out}}^{\mathrm{t}}/\partial y)_\tau$, the net work output optimized pressure ratio $\pi_{\text{max,w}}$ can be determined:

$$\left(\frac{\partial w^{\text{t}}_{\text{out}}}{\partial y}\right)_\tau = -c_\text{p}T_1\left(-1+\frac{\tau}{y^2}\right) \stackrel{!}{=} 0$$

$$\Rightarrow \pi^{1-1/\gamma}_{\text{max,w}} = \tau^{1/2} \Leftrightarrow \pi_{\text{max,w}} = \tau^{\gamma/2(\gamma-1)} \Leftrightarrow p_{2,3} = p_{\text{am}}\tau^{\gamma/2(\gamma-1)} \tag{4.33}$$

$$\Rightarrow \eta_1 = 1 - \sqrt{\frac{T_1}{T_3}}.$$

Using equation (A.20) from the appendix, the temperatures at the turbine and compressor outlets can be compared for this net work output maximizing pressure ratio $\pi_{\text{max,w}}$:

$$\begin{aligned} T_2 &= T_1\pi^{1-1/\gamma} = T_1\tau^{1/2} \\ T_4 &= \frac{T_3}{\pi^{1-1/\gamma}} = \frac{T_3}{\tau^{1/2}} \\ \Rightarrow \frac{T_4}{T_2} &= \frac{\tau}{\pi^{2-2/\gamma}} \underbrace{=}_{\text{eq. (4.33)}} 1. \end{aligned} \tag{4.34}$$

At the pressure ratio $\pi_{\text{max,w}}$ leading to a maximal net output of technical work, the temperatures T_2 at the compressor outlet and T_4 at the turbine outlet are identical. Note that the first-law efficiency is not maximal at these parameters, as it increases with an increasing pressure ratio, as evident from equation (4.30). This is, however, of limited importance, because the power output per unit mass decreases for $\pi > \pi_{\text{max,w}}$, and the gas turbine would have to run faster and faster to still maintain a significant power output with $T_2 \rightarrow T_3$. Under such operating conditions one faces increasing mechanical losses due to the fast operation. Moreover, larger mass-flow rates would be required, leading to an overall increase in the turbine size which is often not desired.

4.2.3 Efficiency of a gas turbine power plant

In this section the first- and second-law efficiencies associated with the different energy conversion processes taking place in a gas turbine power plant following the Joule–Brayton cycle are assessed. The considerations are analogous to the efficiency considerations undertaken in Section 4.1.2 for a steam power plant. An example calculation using realistic values will then be given in Section 4.2.4.

Combustion: Similarly to the case of a steam power plant, the conversion of the fuels into hot flue gas consumes a part of the exergy of the fuels while leading to no energy losses. This will be discussed in detail in Chapter 6. The first-law efficiency of the combustion process consequently is $\eta_{1,\mathrm{CC}} = 1$, and the second-law efficiency is lower ($\eta_2 \approx 0.7$). The combustion process is the only energy conversion process outside the Joule–Brayton process, as the flue gas is the working fluid itself.

Turbine and compressor: The mechanical losses in turbine and compressor are accounted for analogously to Section 4.1.2. Again, identical first- and second-law efficiencies for the mechanical energy conversion processes are obtained:

$$\eta_{1,\mathrm{mech}} = \eta_{2,\mathrm{mech}} =: \eta_{\mathrm{mech}} = \frac{w^{t}_{\mathrm{out}}}{-({}_1w^t_2 +{}_3 w^t_4)} = \frac{-({}_1w^t_2/\eta_{\mathrm{comp}} +{}_3 w^t_4 \cdot \eta_{\mathrm{turb}})}{-({}_1w^t_2 +{}_3 w^t_4)}, \tag{4.35}$$

where η_{comp} and η_{turb} account for the irreversibilities in compressor and turbine, respectively. They are defined by the ratio of compressor work input in the ideal case and the actual work necessary to run the compressor, and the ratio between maximal turbine work and actual turbine work, respectively.

Exhaust gas: A major reason for efficiency losses in a gas turbine power plant is the elevated temperature of the exhaust gas $T = T_4 \gg T_{\mathrm{am}}$. The exergy losses by the exhaust gas release lead to a net exergy input lower than the exergy content of the driving heat ${}_2q_3$. The ratio between the net exergy input and the total exergy content of the driving heat then marks the second-law efficiency of an ideal gas turbine following a Joule-Brayton cycle:

$$\Rightarrow \eta_{2,\mathrm{ideal}} = \frac{\psi_{\mathrm{in,net}}}{{}_2\psi_3} = 1 - \frac{{}_4\psi_1}{{}_2\psi_3}. \tag{4.36}$$

An exergy flow diagram of the energy and exergy conversion processes is shown in Figure 4.14.

As evident from Figure 4.14, the main energy and exergy losses occur in the same step, namely due to the release of the hot exhaust gas. The overall first- and second-law efficiencies for the gas turbine power plant can then be written as

$$\begin{aligned}\eta_1 &= \frac{w^t_{\mathrm{out}}}{h_{\mathrm{in,fuel}}} = \eta_{1,\mathrm{comb}} \cdot \frac{w^t_{\mathrm{out}}}{{}_2q_3} = \eta_{1,\mathrm{comb}} \cdot \eta_{\mathrm{mech}} \cdot \left(1 - \pi^{1/\gamma - 1}\right)\\ &= \eta_{\mathrm{mech}} \cdot \left(1 - \pi^{1/\gamma-1}\right) \qquad (4.37)\\ \eta_2 &= \frac{w^t_{\mathrm{out}}}{\psi_{\mathrm{in,fuel}}} = \eta_{2,\mathrm{comb}} \cdot \frac{w^t_{\mathrm{out}}}{{}_2\psi_3} = \eta_{2,\mathrm{comb}} \cdot \eta_{\mathrm{mech}} \cdot \eta_{2,\mathrm{ideal}}.\end{aligned}$$

4.2.4 Example calculation

As an example a gas turbine power plant operated with realistic parameters is considered. State '1' at the compressor inlet is the ambient state, and the maximal temperature is $T_3 = 800\,^{\circ}\mathrm{C} = 1073$ K. While in reality the adiabatic exponent γ will change

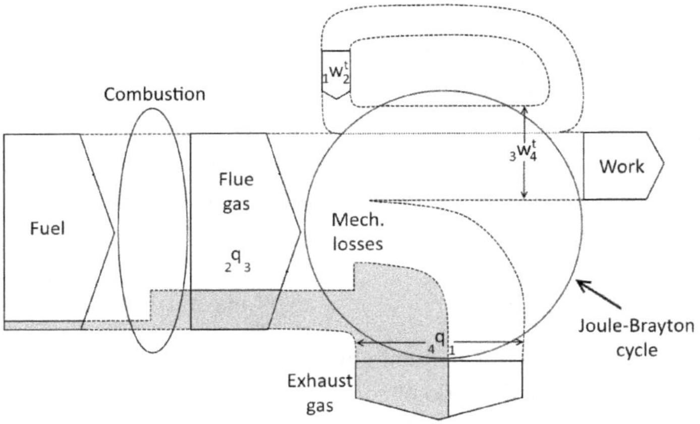

Fig. 4.14: Energy and exergy flow diagram for the energy conversion processes in a gas turbine power plant. The *ellipsoids* represent energy conversion processes, and the *arrows* represent transferred energy, with the *upper white fraction of the arrows* representing the transferred exergy.

slightly in the temperature range considered, this change is neglected here, and a constant value of $\gamma = 1.4 = 7/5$ is assumed. For a derivation of this value see Section A.2 of the appendix. Furthermore, an operation of the power plant at the work output optimized pressure ratio $\pi_{\text{max,w}}$ is assumed. For a mechanical efficiency of $\eta_{\text{mech}} = 0.8$ the overall first-law efficiency can then already be given using equations (4.33) and (4.37):

$$\eta_1 = 0.8 \cdot \left(1 - \sqrt{\frac{298}{1073}}\right) = 0.38. \tag{4.38}$$

For the calculation of the second-law efficiency, $\pi_{\text{max,w}}$ and $T_2 = T_4$ have to be determined in a first step using equations (4.33) and (4.34):

$$\begin{aligned} \pi_{\text{max,w}} &= \left(\frac{1073}{298}\right)^{7/4} = 9.4 \\ \Rightarrow T_2 = T_4 &= 298 \cdot \left(\frac{1073}{298}\right)^{1/2} \text{ K} = 565 \text{ K}. \end{aligned} \tag{4.39}$$

With this and equation (3.22), the exergy transfers at the two virtual heat exchangers ${}_2\psi_3$ and ${}_4\psi_1$ can be calculated using a constant heat capacity of $c_p = 7/2R = 1.01$ J/(gK), with the gas constant R:

$$\begin{aligned} {}_2\psi_3 &= \int_{565\text{ K}}^{1073\text{ K}} 1.01\text{ J/gK} \cdot \left(1 - \frac{298\text{ K}}{T}\right) dT = 320\text{ J/g} \\ {}_4\psi_1 &= \int_{298\text{ K}}^{565\text{ K}} 1.01\text{ J/gK} \cdot \left(1 - \frac{298\text{ K}}{T}\right) dT = 77.1\text{ J/g} \\ \Rightarrow \quad \eta_{2,\text{ideal}} &= 0.76. \end{aligned} \tag{4.40}$$

Consequently about a fourth of the exergy of the input heat is rejected with the exhaust gas. The overall second-law efficiency is then given by

$$\eta_2 = 0.7 \cdot 0.8 \cdot 0.76 = 0.43. \tag{4.41}$$

If the combustion process is factored out and the hot flue gas is taken as the energy input, the second-law efficiency increases to $\eta_2 = 0.61$. In this case it is dominated by the losses due to the elevated exhaust gas temperature. These values are already realistic for real gas turbine power plants. It must be noted, however, that the assumption of a constant c_p and consequently a constant γ is not only violated due to the properties of air itself, but also due to the fact that the flue gas contains CO_2 and water. For more realistic assumptions of c_p the calculated efficiencies would be slightly lower.

4.2.5 Intercooling

In a real gas-fired power plant several modifications can be applied to further increase the first- and second-law efficiencies. The two modifications discussed in this and the following section address two problems arising in a gas-fired power plant, namely the high amount of technical work needed for the compression and the high temperature of the exhaust gas. The former problem is mitigated by a modification in the Joule–Brayton cycle called **intercooling.** For the latter problem, two different approaches are taken, depending on the design of the gas turbine. For a stand-alone gas turbine, which for example is used when fast load changes have to be realized, regenerative heating similar to the feed water heating in a steam power plant discussed in Section 4.1.4 is implemented. Note that in this case T_4 has to be larger than T_2. Alternatively, the hot exhaust gas can be used to drive a steam power plant in so-called **combined cycle power plants.**

To reduce the work required for the compressor the process is typically split into two parts. The working fluid is first compressed to a pressure level p_{1a}, where it is cooled down by the ambient air in a closed heat exchanger and afterward compressed to the upper pressure level p_2 in a second step. The Ts- and pv-diagrams for the thus modified Joule–Brayton process are shown in Figure 4.15.

The technical work saved by the intercooling can be directly seen as the marked area in the pV-diagram in Figure 4.15. It becomes also evident that the compression process is approaching an isothermal compression, which is theoretically the optimal path in terms of efficiency. For the pressure level p_{1a} an optimal choice with respect to the net output technical work exists, which can be seen by first calculating and then minimizing the technical work for the two-step process using equations (3.14) and (A.19), under the assumption that the working fluid is cooled down to ambient

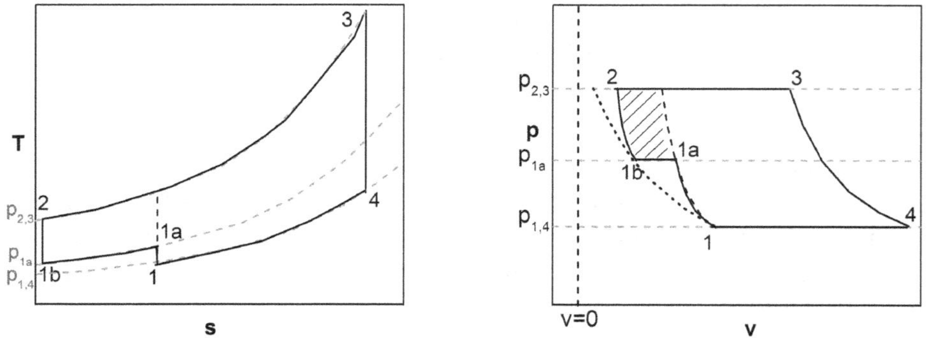

Fig. 4.15: Ts- and pv-diagrams for a Joule–Brayton cycle with intercooling (*solid lines*) together with the unmodified process (*dashed lines*). The marked area in the pv-diagram corresponds to the technical work saved in the compression process with intercooling. The *dotted line* in the pv-diagram represents the isotherm at $T = T_{\mathrm{am}}$.

temperature T_{am} at the point 1b:

$$
\begin{aligned}
w^{\mathrm{t}}_{\mathrm{ic}} &= \int_{p_1}^{p_{1a}} 1/\varrho(p)dp + \int_{p_{1a}}^{p_2} 1/\varrho(p)dp \\
&= \frac{\gamma}{\gamma-1}\left[\frac{p_1^{1/\gamma}}{\varrho_1}\left(p_{1a}^{1-1/\gamma} - p_1^{1-1/\gamma}\right) + \frac{p_{1b}^{1/\gamma}}{\varrho_{1b}}\left(p_2^{1-1/\gamma} - p_{1b}^{1-1/\gamma}\right)\right] \\
&= \frac{\gamma}{\gamma-1}\left[\frac{p_1}{\varrho_1}\left(\left(\frac{p_{1a}}{p_1}\right)^{1-1/\gamma} - 1\right) + \frac{p_{1b}}{\varrho_{1b}}\left(\left(\frac{p_2}{p_{1a}}\right)^{1-1/\gamma} - 1\right)\right] \\
&= \frac{\gamma}{\gamma-1} RT_{\mathrm{am}}\left[\left(\frac{p_{1a}}{p_1}\right)^{1-1/\gamma} + \left(\frac{p_2}{p_{1a}}\right)^{1-1/\gamma} - 2\right]
\end{aligned}
\tag{4.42}
$$

Note that the gas constant R is here a mass-specific quantity and not, as is more common, a mole-specific quantity. For the first derivative of $w^{\mathrm{t}}_{\mathrm{ic}}$ this leads to

$$
\begin{aligned}
\frac{dw^{\mathrm{t}}_{\mathrm{ic}}}{dp_{1a}} &= \frac{\gamma}{\gamma-1} RT_{\mathrm{am}} \frac{\gamma-1}{\gamma}\left[p_1^{-1}\left(\frac{p_{1a}}{p_1}\right)^{-1/\gamma} - p_2/p_{1a}^2 \cdot \left(\frac{p_2}{p_{1a}}\right)^{-1/\gamma}\right] \stackrel{!}{=} 0 \\
\Rightarrow 0 &= p_1^{-(1-1/\gamma)} p_{1a}^{-1/\gamma} - p_2^{1-1/\gamma} p_{1a}^{1/\gamma-2} \\
\Leftrightarrow 0 &= p_{1a}^{-1/\gamma} - \pi^{1-1/\gamma} p_1^{2-2/\gamma} p_{1a}^{1/\gamma-2} \\
\Leftrightarrow p_{1a}^{2-2/\gamma} &= \left(\pi p_1^2\right)^{1-1/\gamma} \\
\Leftrightarrow \left(\frac{p_{1a}}{p_1}\right)^2 &= \pi,
\end{aligned}
\tag{4.43}
$$

where again π denotes the ratio between the maximal and the minimal pressures in the system. Thus, the choice for the intermediate pressure level p_{1a} maximizing the net work output is the geometrical average of p_1 and p_2, i.e. $p_{1a} = \sqrt{p_1 p_2}$. For a multistep intercooling process with j steps this can be generalized, and the pressure of the i-th step is then given by

$$\left(\frac{p_{1,i}}{p_{am}}\right) = \pi^{i/j}. \tag{4.44}$$

4.2.6 Combined cycle power plants (CCPP)

The major improvement possible for gas turbine power plants is to use the exhaust gas to drive an independent steam power plant, as depicted in Figure 4.16.

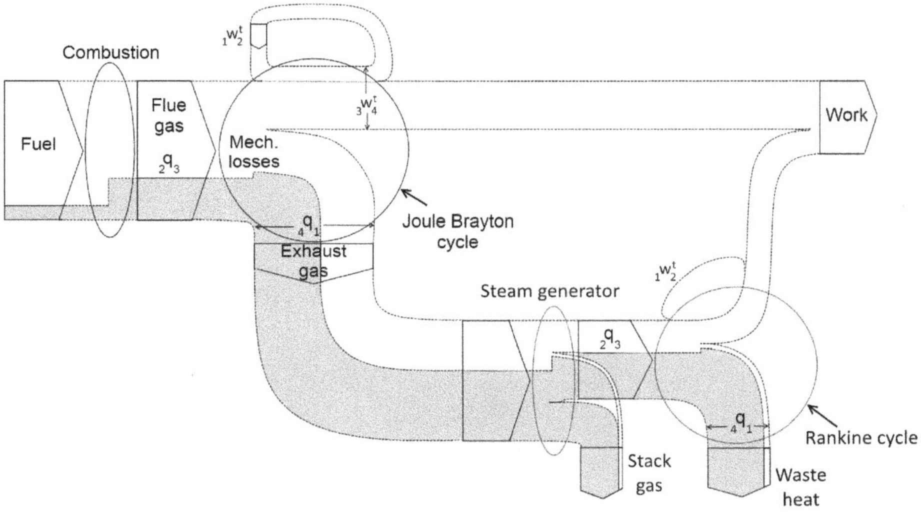

Fig. 4.16: Energy and exergy flow diagram of a combined cycle power plant (CCPP).

In such a power plant the exhaust gas at elevated temperature T_4 of the gas turbine power plant plays the role of the heat source in the steam power plant. In all other parts, the heat engine and cooling parts, of the steam power plant are identical to Section 4.1. The first- and second-law efficiencies of the CCPP are simply given by the sum of the technical work outputs of the two cycles divided by the total input of energy or exergy, respectively:

$$\begin{aligned} \eta_1 &= \frac{-(w^t_{gas} + w^t_{steam})}{h_{in,fuel}} \\ \eta_2 &= \frac{-(w^t_{gas} + w^t_{steam})}{\psi_{in,fuel}}. \end{aligned} \tag{4.45}$$

Typically the work output of the gas turbine exceeds the work output of the steam turbine by a factor of 2. In this case the first- and second-law efficiencies are given by 3/2 of the values calculated for the gas turbine power plant in equations (4.38) and (4.40), leading to values of $\eta_{1,\mathrm{CCPP}} = 0.57$ and $\eta_{2,\mathrm{CCPP}} = 0.65$. If the exergy loss due to the combustion process itself is neglected the latter value rises to $\eta_{2,\mathrm{CCPP}} = 0.86$.

5 Electrical exergy

So far, the end product of all the processes considered was mechanical work provided by a turbine. Now a look is taken at the typically desired form of final energy delivered by these processes, namely electrical exergy. To do this, in a first step the voltage sources are treated as black boxes, and the properties of the electrons in the two electrical leads of these black boxes are the subject of interest. This means that this chapter is concerned with the ensemble of electrons in conductors as the working medium. To investigate the properties of this medium, some basics from the field of solid state physics have to be considered. More details on particle ensembles and solid state physics can be found in Appendix B. In a second step, in this and the following chapters, a look into these black boxes, the voltage sources, is taken, and the underlying driving forces are identified.

5.1 The electrochemical potential

To understand the behavior of the electron ensemble, the quantum nature of the electrons has to be taken into account. Note that it is usually not helpful to envision the electrons as single particles, but that instead the notion of an ensemble with distinct properties as a whole is much more useful. The properties of this ensemble are then determined by two main characteristics: first, by the composition and structure of the solid, and, second, by the quantum mechanical interactions between the individual electrons. As for the former, it sets the background of electronic states, which can then either be occupied or unoccupied, and can be characterized according to their energy. Furthermore, the total number of free electrons is set by the elements constituting the solid. As for the latter, electrons are fermions. This means that they are subject to the **Pauli exclusion principle**, which states that no two fermions can occupy an identical quantum mechanical state. In a metal a number of free mobile electrons comparable to the number of atoms in the metal is available. The energetic equilibrium state of the ensemble of all electrons can be found by minimizing the Gibbs free energy G of the electron ensemble while fulfilling the Pauli exclusion principle. The resulting distribution function is the **Fermi–Dirac distribution** (for details see Section B.1.2 of the appendix):

$$f(E,T) = \frac{1}{1+e^{(E-\tilde{\mu})/kT}}, \tag{5.1}$$

where $k = 8.62 \cdot 10^{-5}$ eV/K is the Boltzmann constant. The Fermi–Dirac distribution $f(E,T)$ gives the probability that an electronic state with the energy E is occupied when the electron ensemble is kept at a temperature T. The quantity $\tilde{\mu}$ in the distribution function is called the **electrochemical potential** of the electron ensemble. It marks the energy at which the occupation probability is 1/2 and is the most important quan-

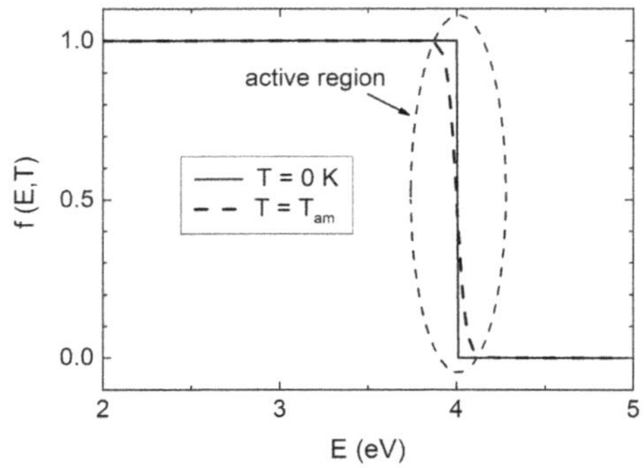

Fig. 5.1: Fermi–Dirac distribution at two different temperatures for $\tilde{\mu}$ = 4 eV.

tity to describe the properties of an ensemble of electrons. Note that $\tilde{\mu}$ is typically in the range of a few eV, while the thermal energy is $kT \approx 25$ meV at room temperature. At $T = T_{\mathrm{am}}$, this leads to a rather steep decline of $f(E, T)$ from 1 to 0 around $\tilde{\mu}$ as shown in Figure 5.1.

Only the electrons at energies where the Fermi–Dirac distribution takes values different from 1 and 0 contribute to the overall properties of the ensemble. Often in solid state physics, the electrochemical potential is called the **Fermi energy** E_{F}. This denotation is not quite correct, as the Fermi energy is defined as the maximal electronic energy at a temperature $T = 0$ K. In reality, however, $\tilde{\mu}$ is slightly temperature dependent. Therefore, the expression

$$\tilde{\mu}(T = 0\,\mathrm{K}) = E_{\mathrm{F}} \tag{5.2}$$

holds, while in general $E_{\mathrm{F}} \neq \tilde{\mu}(T)$. Still, the Fermi energy is a good approximation for the electrochemical potential in metals, due to the relatively small variation of $\tilde{\mu}(T)$ with T. In semiconductors, however, the variation can be larger, and the term “electrochemical potential” rather than “Fermi energy” should always be used. The electrochemical potential of any charged particle with charge $q := ze$, with the charge number z ($z > 0$ for positively and $z < 0$ for negatively charged particles) and the elementary charge e, can be given for every point in space and can be decomposed into two contributions:

$$\tilde{\mu}(\vec{x}) = \mu(\vec{x}) + q\varphi(\vec{x}). \tag{5.3}$$

Here, $\mu(\vec{x})$ is the spatial distribution of the **chemical potential**, i.e. the energy necessary to change the configuration of the particle ensemble by adding or subtracting a particle at position $\vec{x}$ and $\varphi(\vec{x})$ stands for the electrostatic potential. For electrons $z = -1$ holds and the electrochemical potential is thus given by

$$\tilde{\mu}(\vec{x}) = \mu(\vec{x}) - e\varphi(\vec{x}). \tag{5.4}$$

When dealing with chemical reactions, it is typically useful to deal with the amounts of substances measured in mole rather than with the number of particles. In these cases the molar chemical and electrochemical potentials μ_m and $\tilde{\mu}_\mathrm{m}$ are used, which represent the changes in free energy when one mol of substance is exchanged with the ensemble and have the unit J/mol:

$$\tilde{\mu}_\mathrm{m} = \mu_\mathrm{m} + zF\varphi . \tag{5.5}$$

Here, $F = N_\mathrm{A} \cdot e = 96485$ C/mol is the **Faraday constant**, which is the charge of one mole of elementary charges measured in Coulomb. If an electron is added from the vacuum to the electron ensemble, the Gibbs free energy of the latter increases by $\tilde{\mu}$, and if an electron is taken from the ensemble and transferred to the vacuum it decreases by $\tilde{\mu}$. As the free energy is the entropy-free part of the energy, at room temperature the exergy exchange by transferring electrons across the borders of the electron ensemble is given by

$$\psi_\mathrm{m} = \pm\tilde{\mu}, \tag{5.6}$$

where the + sign is valid for the addition of an electron to the ensemble and the – sign for the subtraction of an electron from the ensemble, in accordance with the sign convention for energy transfers given in Section 2.1.1.

5.1.1 Interfaces

In equilibrium, the electrochemical potential of all particles from which the matter is formed is constant in space, as this corresponds to the state of minimal free energy. Whenever two material phases are brought into contact which both contain a species α that can cross the phase boundary, the species will, in general, have different electrochemical potentials $\tilde{\mu}_1^\alpha$ and $\tilde{\mu}_2^\alpha$ before contact. Note that, whenever a species other than electrons are referred to, the type of species is given by a superscript. Upon contact, the free energy of the system is lowered by adapting the electrochemical potentials of species α on both sides. In the case of uncharged particles, the electrochemical potential is only determined by the chemical potential, and an alignment of the values of the latter on both sides of the interface leads to a spatially uniform activity of the particles in the entire system. The picture is very different for charged particles. Even relatively few exchanges of charged particles across the interface lead to a charging of the materials on both sides of the interface and a corresponding shift in the electrostatic potentials which is relatively large, i.e. in the order of the typical energies of interest in the system. As only very few charge carriers, compared with the total amount of available charge carriers, are exchanged, the chemical potentials of the species remain virtually unchanged. Thus, both materials are charged in the contact region, which gives rise to some drop of the electrostatic potential across the interfacial region, compensating the difference in the chemical potentials of the species in the two phases. It is this charging which leads to the adaptation of the electrochemical potentials while the chemical potentials in the bulks of the two materials remain unaffected:

$$0 = \tilde{\mu}_2 - \tilde{\mu}_1^{\alpha} = \mu_2^{\alpha} - \mu_1^{\alpha} + q(\varphi_2 - \varphi_1)$$
$$\Rightarrow \mu_2^{\alpha} - \mu_1^{\alpha} = -q(\varphi_2 - \varphi_1) =: -qU_c. \tag{5.7}$$

Here U_c is the so-called contact voltage. As the contact voltage U_c is not linked to a change in the electrochemical potential, it can neither be measured with a voltmeter nor used to gain energy from it. The opposite charges in both materials are located close to the interface as shown in Figure 5.2.

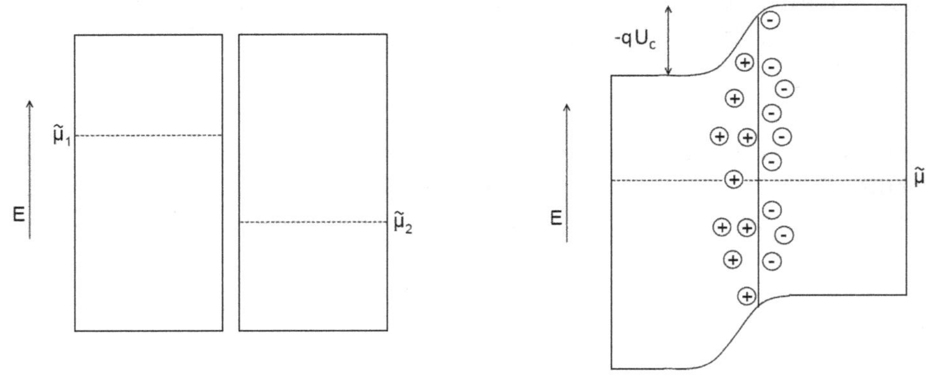

Fig. 5.2: Occurrence of a space charge region *(right)* upon contacting two materials with different electrochemical potentials $\tilde{\mu}_1$ and $\tilde{\mu}_2$ *(left)*. The space charge region extends less far into the material, with the higher conductivity here assumed to be on the *right-hand side*.

The interfacial region where these charges are located is called **space charge region** or **space charge layer**. The excess charges have opposite sign and equal magnitude in the two phases, and their extension into the respective materials is inversely dependent on their respective conductivities. In the case of metals this extension is negligible, as the electronic conductivity is very high, resulting in a pure surface charge. A redox species in an electrolyte solution can exchange electrons with a conductive electrode. Hence, in equilibrium the electrochemical potential of the electrons in the electrode and on the redox-pair are identical and, consequently, a space charge region called **double layer** forms. In an electrolyte the charge carrier density is much lower than in a metal, and the extension of the space charge layer has a typical length scale of ≈ 1 nm. For semiconductors the charges in the a space charge region are charged doping atoms. As these are sitting at fixed positions, the extension of the space charge region is determined by their volume density and reaches values of several 1–10 μm. The alignment of the electrochemical potentials at both sides of the interface is only possible for those species that can cross it. For other species a difference in the electrochemical potentials on both sides remains.

5.1.2 Electrical currents

An electrical current density can be seen as a charge density ϱ_{el} moving with a drift velocity $\vec{v}_{\mathrm{d}}$:

$$\vec{j}(\vec{x}) = \varrho_{\mathrm{el}}(\vec{x}) \cdot \vec{v}_{\mathrm{d}}(\vec{x}) = n(\vec{x}) \cdot q \cdot \vec{v}_{\mathrm{d}}(\vec{x}) \tag{5.8}$$

Here $n(\vec{x})$ denotes the volume density of the ensemble of charge carriers with the charge $q = ze$ at position $\vec{x}$. These currents are not accompanied by local changes in the charge density, in the same way that water currents are not accompanied by changes in the volume density. They are an independent property of the charge carrier ensemble and require a constant emission of charge carriers at one end of the solid and a corresponding constant absorption of them at the other end. As long as no electrical capacities are present, there is no accumulation of charge carriers. Gradients in the electrochemical potential give rise to particle currents, as the free energy of the electron ensemble can be reduced by transferring particles from points with higher electrochemical potential to points with lower electrochemical potential. The force on the individual particles due to the gradient in the electrochemical potential is given by

$$\vec{f} = -\nabla\tilde{\mu}. \tag{5.9}$$

At a given driving force, the drift velocity is determined by scattering processes, which can be well captured by the material and charge carrier type dependent mobility b, measured in $\mathrm{m^2V^{-1}s^{-1}}$:

$$\vec{v}_{\mathrm{d}}(\vec{x}) =: \frac{b}{q} \cdot \vec{f} = -\frac{b(\vec{x})}{q} \cdot \nabla\tilde{\mu}(\vec{x}). \tag{5.10}$$

The mobility is independent of the charge carrier density and is a measure of the quality of the conducting material. Note that in textbooks the mobility is typically denoted μ. This would be an inconvenient choice here, due to the same symbol for the chemical potential. Combining equations (5.10) and (5.8), a generalized version of Ohm's law is obtained:

$$\vec{j}(\vec{x}) = n(\vec{x}) \cdot q \cdot \vec{v}_{\mathrm{d}}(\vec{x}) = -n(\vec{x}) \cdot b(\vec{x}) \cdot \nabla\tilde{\mu}(\vec{x}), \tag{5.11}$$

with the particle density $n(\vec{x}) = \varrho_{\mathrm{el}}(\vec{x})/q$. Making use of equation (5.3), the total current can be decomposed into two parts, first a diffusion current due to a spatial variation of the chemical potential, and, second, a drift current due to an electric field $\vec{E} = -\nabla\varphi$:

$$\begin{aligned} \vec{j}(\vec{x}) &= -n(\vec{x}) \cdot b(\vec{x}) \cdot (\nabla\mu(\vec{x}) + q\nabla\varphi(\vec{x})) \\ &= \underbrace{-n(\vec{x}) \cdot b(\vec{x}) \cdot \nabla\mu(\vec{x})}_{\text{diffusion}} + \underbrace{n(\vec{x}) \cdot b(\vec{x}) \cdot q \cdot \vec{E}(\vec{x})}_{\text{drift}}. \end{aligned} \tag{5.12}$$

In case of a spatially constant chemical potential, the diffusion current is zero, and equation (5.11) reduces to the typical form of Ohm's law, stated in terms of the electrical conductivity $\sigma(\vec{x}) = \varrho_{\mathrm{el}}(\vec{x})b(\vec{x})$:

$$\vec{j}(\vec{x}) = n(\vec{x}) \cdot b(\vec{x}) \cdot q \cdot \vec{E}(\vec{x}) = \varrho_{\mathrm{el}}(\vec{x}) \cdot b(\vec{x}) \cdot \vec{E}(\vec{x}) = \sigma(\vec{x})\vec{E}(\vec{x}). \tag{5.13}$$

5.2 Voltage sources

The defining property of a voltage source is that the electron ensembles in the electrical leads from a voltage source have to have different electrochemical potentials. Only in this case is a measurable and usable external voltage present, and only then can energy be gained by transferring electrons from the lead with the higher electrochemical potential to the lead with the lower electrochemical potential. As the leads are typically made from the same material, normally copper, and are held at the same temperature $T = T_{am}$, the chemical potential of the electrons is identical in both leads, and any difference in the electrochemical potentials is consequently an electrical potential difference. The energy E provided by the voltage source upon the transfer of an electron from lead 1 into lead 2 through a consumer is given by the difference in the respective electrochemical potentials:

$$E = \tilde{\mu}_2 - \tilde{\mu}_1 = e\,(\varphi_1 - \varphi_2) =: -eU. \tag{5.14}$$

This, together with the exergy exchange by charge carrier transfers in electron ensembles given in equation (5.6), already leads to the following important conclusion:

Electrical energy is pure exergy if the leads (made from the same material) are kept at $T = T_{am}$ and can be approximated as pure exergy at elevated temperatures $T > T_{am}$. !

For any practical purpose electrical energy can thus be treated as pure exergy. Any voltage source can be seen as an open system, with the electron ensemble as the working medium. For a given voltage source, the current, i.e. the flow rate of the working medium, is then set by the driving force for the current provided by the voltage source in interaction with a given external consumer. Note that the expression for the electrical power $P = UI$ ties in nicely with the view of a voltage source as an open system component. As for all open systems, the total power input into the working fluid is given by the product of the particle current and the change in the particle energy upon crossing the system (see Chapter 3). In general, the voltage provided by the voltage source is then dependent on this current, which can be used as a parameter in the description of the voltage sources and characteristic curves, current voltage (*UI*-)characteristics, can be given for voltage sources. In the steady flow equilibrium with a given current I, the voltage source provides a different electrochemical potential of the electrons at both of its electrical connections according to its *UI*-characteristic. As in equilibrium the electrochemical potential of the electrons is expected to be identical at each point of the system, by design any voltage source has to contain parts, e.g. semipermeable membranes, which inhibit the unhindered flow within the voltage source of some of the species that determine the electrostatic potential of the leads.

Any voltage source has a built-in asymmetry inhibiting the adjustment of the electrochemical potentials of the potential determining species. To understand why a voltage source works, the origin of the asymmetry has to be identified.

In the final sections of this chapter and in the following chapters, a variety of such asymmetries will be discussed. As was done in the previous chapters, the analysis of voltage sources will be done in two steps. First, a characterization of the ideal conversion processes leading to the open circuit voltage U^{oc}, i.e. the voltage for $I = 0$, is given, and, second, an analysis of the losses due to the necessary driving forces as expressed in *UI*-curves is performed. Note that it is not helpful to envisage the positive and negative poles of a voltage source as the result of an electron depletion and an electron enrichment, respectively. This picture is mostly unhelpful and often misleading. While surface charges are typically present and do indeed set the potentials of the leads, they are a property of the voltage source itself. Thus, the *UI*-characteristic of a voltage source is an intrinsic property of the voltage source, not of the surface charges.

5.3 Generators

As a first voltage source, the electrical **generator** which converts mechanical technical work into electrical energy is described. An AC generator typically consists of a conducting wire arranged in several coils, which are then exposed to a varying magnetic field typically realized by a turnable magnet placed in their center, as shown in Figure 5.3. Generators used in power plants work different in detail, but the overall concepts are identical to the ones discussed in this section.

The working medium in a generator is the ensemble of charge carriers in the conducting wire. As described above, the only way to change the free energy of this working medium is to change its electrochemical potential $\tilde{\mu}(\vec{x})$. This can be achieved by rotating the magnet and thus changing the total magnetic flux Φ through the coils, giving rise to a difference in the electrostatic potential $\Delta\varphi$ between the ends of the wire according to the law of induction:

$$\dot{\Phi} = -\Delta\varphi = \varphi_{\mathrm{in}} - \varphi_{\mathrm{out}}. \tag{5.15}$$

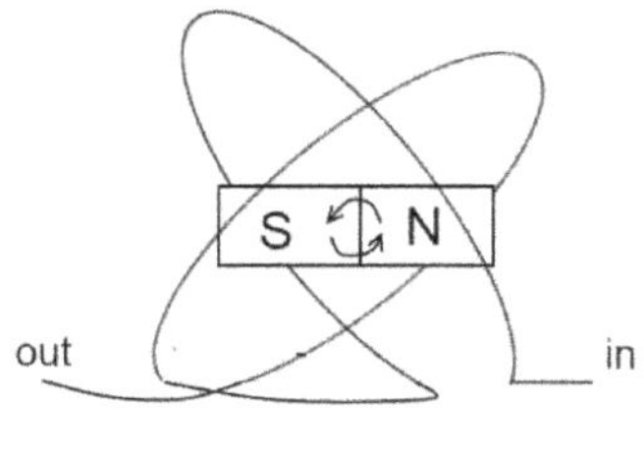

Fig. 5.3: Schematic of a generator with two windings.

As the chemical potential of the electrons in the conductor is unaffected, the total difference in the electrochemical potential $\tilde{\mu}$ between the contacts of the wire has the identical absolute value as the change in the electrostatic energy:

$$\Delta\tilde{\mu} = -e\Delta\varphi. \tag{5.16}$$

The rotational direction of the magnet introduces the asymmetry underlying the function of the generator as a voltage source. For a given rotational direction, a current in one direction leads to an energy output, and a current in the other direction requires an energy input.

5.3.1 Electrical power output

For the further considerations the current through the generator has to be considered. This current is not determined by the negligible resistance of the wire in the generator itself, but by the consumer. Thus, the generator is treated as an open system component, where the flux of the working fluid, i.e. the electron ensemble, is externally set. In the following derivation AC voltage and current will be used, which is natural for a generator but not essential for the arguments, and all considerations are also applicable for DC voltages and currents.

For a given magnetic field $\vec{B}$, the voltage produced by the magnet rotating with the frequency ω is given by the angle ωt between the area encircled by the wire and $\vec{B}$:

$$U(t) = -\dot{\Phi} = -\vec{B}\cdot\dot{\vec{A}} = -BA\frac{d}{dt}\cos(\omega t) = BA\omega\sin(\omega t) =: |\widehat{U}|\cos(\varphi'), \tag{5.17}$$

where $|\widehat{U}| = BA\omega$ is the peak voltage output, and the angle $\varphi' = \omega t + \pi/2$ is introduced. The current response of a consumer to this applied AC voltage is determined by its complex **impedance** $Z = R + i(\omega L + 1/\omega C)$ (R, L, and C are the **resistance**, **inductance**, and **capacitance** of the consumer, respectively), which affects both the amplitude and the phase of the current. For easier calculation the complex numbers $\widehat{U}(t) = |\widehat{U}|e^{i\varphi'}$ and $\widehat{I} = |\widehat{I}|e^{i\psi'}$ are used, where $Re\left(\widehat{U}(t)\right) = U(t)$ and $Re\left(\widehat{I}(t)\right) = I(t)$. For the current the following expression is then found:

$$\begin{aligned} I(t) &= Re\left(\widehat{I}(t)\right) = Re\left(\frac{\widehat{U}(t)}{Z}\right) = Re\left(\frac{\widehat{U}(t)Z^*}{|Z|^2}\right) \\ &= Re\left(\frac{1}{|Z|^2}|\widehat{U}|R\cdot e^{i\varphi'} + \left(|\widehat{U}|\,(\omega L + 1/\omega C)\right)e^{i(\varphi'-\pi/2)}\right) \\ &= \frac{|\widehat{U}|R}{|Z|^2}\cos(\varphi') + \frac{|\widehat{U}|\,(\omega L + 1/\omega C)}{|Z|^2}\sin(\varphi'), \end{aligned} \tag{5.18}$$

where Z^* denotes the complex conjugate of Z. The average electrical power provided by the generator $\overline{P_{\text{el}}}$ thus becomes

$$\begin{aligned}\overline{P_{\text{el}}} &= \overline{U(t)I(t)}\\ &= \frac{1}{2\pi}\int_0^{2\pi}\left(\frac{|\widehat{U}|^2 R}{|Z|^2}\cos^2(\varphi') + \frac{|\widehat{U}|^2\,(\omega L + 1/\omega C)}{|Z|^2}\cos(\varphi')\sin(\varphi')\right)d\varphi'\\ &= \frac{1}{2}\frac{|\widehat{U}|^2 R}{|Z|^2} = \frac{1}{2}\frac{B^2A^2\omega^2 R}{|Z|^2},\end{aligned} \tag{5.19}$$

where the fact that the integral over a full period of $\sin(\varphi')\cos(\varphi')$ is equal to zero is used. For the last step, in equation (5.19) $|\widehat{U}|$ is taken from equation (5.17).

5.3.2 Mechanical power input

To determine the input of mechanical technical work, the torque $\vec{M}(t)$ required to turn the magnet in the magnetic field generated by the current in the coils has to be calculated. The magnetic moment $\vec{m}(t)$ of the coil with current I is given by

$$\vec{m}(t) = I(t)\vec{A} = \left(\frac{|\widehat{U}|R}{|Z|^2}\cos(\varphi') + \frac{|\widehat{U}|\,(\omega L + 1/\omega C)}{|Z|^2}\sin(\varphi')\right)\vec{A}. \tag{5.20}$$

The torque $\vec{M}$ required to turn the magnet is then given by the vector product of the magnetic moment $\vec{m}$ from the coils and the magnetic field of the permanent magnet $\vec{B}$:

$$|\vec{M}(t)| = |\vec{m}(t)\times\vec{B}| = \left(\frac{|\widehat{U}|R}{|Z|^2}\cos(\varphi') + \frac{|\widehat{U}|\,(\omega L + 1/\omega C)}{|Z|^2}\sin(\varphi')\right)AB\sin(\omega t). \tag{5.21}$$

For the average technical work input per unit time $\overline{P_{\text{mech}}}$ this leads to

$$\begin{aligned}\overline{P_{\text{mech}}} &= \overline{\vec{M}(t)\cdot\vec{\omega}} = \overline{|M(t)|\omega}\\ &= \frac{\omega}{T}\int_0^{T}\left(\left(\frac{|\widehat{U}|R}{|Z|^2}\sin(\omega t) - \frac{|\widehat{U}|\,(\omega L + 1/\omega C)}{|Z|^2}\cos(\omega t)\right)AB\sin(\omega t)\right)dt\\ &= \frac{1}{2}\frac{|\widehat{U}|RAB\omega}{|Z|^2} = \frac{1}{2}\frac{B^2A^2\omega^2 R}{|Z|^2} = \overline{P_{\text{el}}}.\end{aligned} \tag{5.22}$$

This means that the average input of mechanical power into the generator and its average output of electrical power are identical. Losses only occur in the form of mechanical losses in the shaft which increase the torque and losses, due to the finite resistance of the conducting wire in the generator, which lead to a smaller U. Real generators reach first- and second-law efficiencies of

$$\eta_1 = \eta_2 \approx 95\,\%. \tag{5.23}$$

5.4 Thermoelectrics

While for many materials the temperature dependence of the electrochemical potential of the electrons is rather small, it can still be used for the direct conversion of thermal energy into electrical energy. This direct conversion is called thermoelectrics. If any piece of matter is heated on one side to a temperature $T = T_\mathrm{h}$ while the other side is kept at a constant temperature $T = T_\mathrm{c}$, a certain voltage can be measured between the two ends. This voltage is called the thermovoltage U_T and can be expressed as the difference in the electrochemical potentials of the dominant charge carrier species α (charge q) between both ends:

$$qU_\mathrm{T} = \tilde{\mu}^\alpha(T_\mathrm{c}) - \tilde{\mu}^\alpha(T_\mathrm{h}). \tag{5.24}$$

The thermovoltage is usually expressed in terms of the material specific Seebeck coefficient $S_\mathrm{Seeb}(T)$:

$$U_\mathrm{T} = \int_{\text{cold end}}^{\text{hot end}} -\frac{1}{q}\nabla\tilde{\mu}^\alpha\, dx = \int_{\text{cold end}}^{\text{hot end}} -\frac{1}{q}\left(\frac{\partial\tilde{\mu}^\alpha}{\partial T}\right)_\mathrm{p} \nabla T dx =: \int_{\text{cold end}}^{\text{hot end}} S_\mathrm{Seeb}(T)\nabla T dx. \tag{5.25}$$

Note here that the measured thermovoltage is not only given by the thermoelectric material but also by the material of the leads, as the contact points are at different temperatures, while the location of the voltmeter is typically at a point with $T = T_\mathrm{am}$ in both leads. This will be neglected in the further discussion, assuming that the Seebeck coefficient of the leads is known and can be subtracted. As apparent from equation (5.25), the Seebeck coefficient is given by

$$S_\mathrm{Seeb} = -\frac{1}{q}\left(\frac{\partial\tilde{\mu}^\alpha}{\partial T}\right)_\mathrm{p} = -\frac{1}{q}\left(\frac{\partial g}{\partial T}\right)_\mathrm{p} = \frac{s}{q}, \tag{5.26}$$

where s is the entropy per particle. For the last identity in equation (5.26) the fact that the free energy of the particle ensemble is only determined by $\tilde{\mu}$ is used. Expressed in thermodynamic quantities, the Seebeck coefficient is thus the entropy per particle divided by its charge.

5.4.1 Seebeck coefficients

In this section the physical origin and order of magnitude of the Seebeck coefficient is discussed for two material classes: metals and doped semiconductors. The latter have much higher Seebeck coefficients in the range of $\approx$ 100 μV/K, while the former show values at least one order of magnitude lower. The basis for the following calculations is the fact that in any conductor, especially in metals and doped semiconductors, the charge carrier density is constant. In bulk materials, i.e. far away from the surfaces of the material, a charge accumulation is not tolerated in a conductor, reflecting the fact that the electron ensemble in a solid is almost perfectly incompressible.

Under the assumption of a constant electron density, in a metal the change of the electrochemical potential of the electron ensemble with temperature can be calculated using the **Sommerfeld expansion** (see Section B.1.4 of the appendix), which is a good approximation if the temperatures are not too high, and can usually be employed in solid state physics:

$$\left(\frac{\partial\tilde{\mu}}{\partial T}\right)_{\mathrm{p}} = -\frac{k^2\pi^2}{3}\frac{D'(\tilde{\mu})}{D(\tilde{\mu})}\cdot T, \tag{5.27}$$

where $D(\tilde{\mu})$ is the density of states at $E = \tilde{\mu}$ and $D'(\tilde{\mu}) = dD(E)/dE|_{E=\tilde{\mu}}$. A high Seebeck coefficient is thus reached when the density of states is rapidly changing at the electrochemical potential, while $D(\tilde{\mu})$ has a relatively low value. In many metals the density of states is given by equation (B.5) from the appendix. The resulting Seebeck coefficient is then given by

$$S_{\mathrm{Seeb}} = -\frac{1}{q}\left(\frac{\partial\tilde{\mu}}{\partial T}\right)_{\mathrm{p}} = \frac{1}{q}\frac{k^2\pi^2}{3}\frac{1}{2\tilde{\mu}}T = \frac{\pi^2 k}{6q}\frac{kT}{\tilde{\mu}}. \tag{5.28}$$

Due to its proportionality to $kT/\tilde{\mu}$, S_{Seeb} is relatively small for most metals ($\approx$ 1 μV/K). Note that in some metals, as, e.g. Nickel, the term $D'(\tilde{\mu}) < 0$, and the resulting Seebeck coefficient is thus positive. Assuming mobile point-like particles, this would lead to the conclusion of positively charged mobile particles rather than electrons. This clearly shows that in a solid the notion of a cloud of individual electrons should be abandoned, and the ensemble with its distinct, collective properties should always be considered instead.

In a semiconductor it is assumed that at room temperature essentially all dopant atoms are ionized, again leading to a constant number of particles. This, together with the distance of the electrochemical potential to the valence band edge given in equation (B.26), leads to the following expression for the Seebeck coefficient for nondegenerately p-doped semiconductors:

$$S_{\mathrm{Seeb}} = -\frac{1}{e}\left(\frac{\partial\tilde{\mu}}{\partial T}\right)_{\mathrm{p}} = \frac{k}{e}\ln\left(\frac{N_{\mathrm{v}}}{n_{\mathrm{h}}}\right), \tag{5.29}$$

where N_v is the effective density of states in the valence band (see equation (B.24) in the appendix) and n_h the density of holes in the valence band. As $k/e = 86\ \mu V/K$ and $\ln(N_v/n_h) \approx 1 - 10$ hold, Seebeck coefficients of $S \approx 100\ \mu V/K$ are found. As shown in equation (5.29), the Seebeck coefficient is positive for a p-type doped material in agreement with the assumption of mobile positively charged holes. For n-type doped materials, a corresponding negative value for S_{Seeb} is found. The sign of the thermovoltage thus offers a fast and easy test to determine the doping type of a given piece of a semiconductor material if the doping type is unknown.

5.4.2 Thermoelectric energy conversion

In a thermoelectric generator in the steady flow equilibrium, a heat flow $\dot{Q}_h$ enters the device at $T = T_h$ and a heat flow $\dot{Q}_c$ leaves the device at $T = T_c$. If the two leads at the end of the thermoelectric generator are linked to a consumer, this heat current is accompanied by an electrical current $j = qnv_d$, as schematically shown in Figure 5.4.

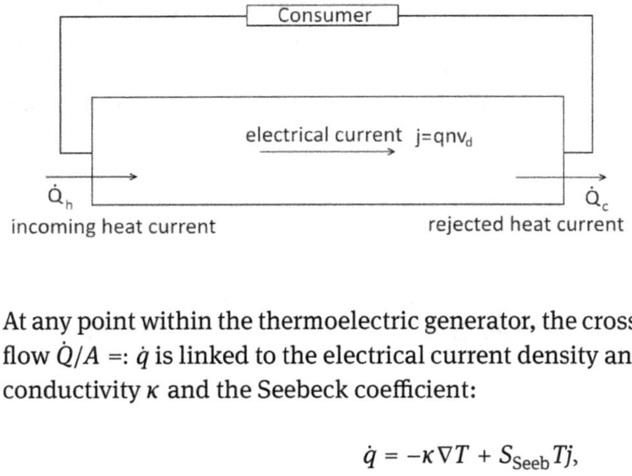

Fig. 5.4: Schematic of a thermoelectric generator connected to an external consumer.

At any point within the thermoelectric generator, the cross sectional area specific heat flow $\dot{Q}/A =: \dot{q}$ is linked to the electrical current density and dependent on the thermal conductivity κ and the Seebeck coefficient:

$$\dot{q} = -\kappa \nabla T + S_{Seeb} Tj, \tag{5.30}$$

where the first term represents the heat current due to the temperature gradient alone, and the second term is the heat current carried by the entropy flow accompanying the charge carrier flow according to $S_{Seeb} Tj = s/q \cdot Tj = sTnv_d$. The second term is thus the entropy per particle multiplied by the motion of the particle ensemble with the drift velocity v_d and temperature. For further considerations of the properties of thermoelectric generators, a relatively small $\Delta T = T_h - T_c$ and a constant electrical conductivity σ, thermal conductivity κ, and also a constant Seebeck coefficient S_{Seeb} are assumed. Furthermore, a linear temperature profile is used, leading to $-\nabla T = \Delta T/d$, where d is the length of the thermoelectric generator from the hot to the cold end. An important quantity typically used in the field of thermoelectrics is the dimensionless figure of merit ZT:

$$ZT = \frac{S_{\text{Seeb}}^2\sigma}{\kappa}T \Rightarrow Z = \frac{S_{\text{Seeb}}^2\sigma}{\kappa}. \tag{5.31}$$

As the internal electrical resistance of the thermoelectric generator R_{int} is an integral part of the following derivation, it will be taken into account from the start. The ratio between the electrical conductivity and the thermal conductivity is especially important for the overall performance, as expressed in the figure of merit ZT. In the steady flow equilibrium the total input heat current $\dot{Q}_{\text{h}}$ entering the hot end of the thermoelectric generator can be written in the following way:

$$\dot{Q}_{\text{h}} = S_{\text{Seeb}}T_{\text{h}}I + \frac{\kappa A}{d}(T_{\text{h}} - T_{\text{c}}) - \frac{1}{2}R_{\text{int}}I^2 = S_{\text{Seeb}}T_{\text{h}}I + \frac{S_{\text{Seeb}}^2}{ZR_{\text{int}}}\Delta T - \frac{1}{2}R_{\text{int}}I^2, \tag{5.32}$$

where the former two terms are taken from equation (5.30), and the last term takes into account the nondirectional heat generation due to the irreversibilities caused by the internal resistance of the thermoelectric generator. The current is set by the thermovoltage and the total resistance of the thermoelectric device and an external consumer with $R_{\text{ext}} =: xR_{\text{int}}$:

$$I = \frac{U_{\text{T}}}{R_{\text{int}} + R_{\text{ext}}} = \frac{U_{\text{T}}}{(1+x)R_{\text{int}}} = \frac{S_{\text{Seeb}}\Delta T}{(1+x)R_{\text{int}}}. \tag{5.33}$$

For the voltage drop U_{ext} across the external consumer and for the corresponding power output P_{el} the following expressions thus hold:

$$\begin{aligned} U_{\text{ext}} &= R_{\text{ext}} \cdot I = U_{\text{T}}\frac{x}{1+x} = \frac{S_{\text{Seeb}}\Delta T}{q} \cdot \frac{x}{1+x} \\ \Rightarrow P_{\text{el}} &= U_{\text{ext}}I = \frac{S_{\text{Seeb}}^2(\Delta T)^2 x}{q(1+x)^2 R_{\text{int}}}. \end{aligned} \tag{5.34}$$

Using equations (5.32)–(5.34), the first-law efficiency can now be written in the following lengthy expression:

$$\begin{aligned} \eta_1 &= \frac{P_{\text{el}}}{\dot{Q}_{\text{h}}} = \frac{UI}{S_{\text{Seeb}}T_{\text{h}}I - 1/2R_{\text{int}}I^2 + S_{\text{Seeb}}^2\Delta T/ZR_{\text{int}}} \\ &= \frac{S_{\text{Seeb}}^2(\Delta T)^2 x/R_{\text{int}}(1+x)^2}{S_{\text{Seeb}}^2 T_{\text{h}}\Delta T/(1+x)R_{\text{int}} - S_{\text{Seeb}}^2(\Delta T)^2/2(1+x)^2R_{\text{int}} + S_{\text{Seeb}}^2\Delta T/ZR_{\text{int}}} \\ &= \frac{\Delta Tx}{T_{\text{h}}(1+x) - \Delta T/2 + (1+x)^2/Z} = \frac{\Delta Tx}{\overline{T}(1+x) + x\Delta T/2 + (1+x)^2/Z}, \end{aligned} \tag{5.35}$$

where $\overline{T} = (T_h + T_c)/2$ is the average temperature in the device. As apparent from equation (5.35), the efficiency is a function of x. It is equal to zero for both $x = 0$ and $x \to \infty$, and thus a global maximum exists. The calculation of this optimal choice of x is omitted here. Its value is $x = \sqrt{1 + Z\overline{T}}$. Inserting this expression for x into equation (5.35) yields for the maximal first- and second-law efficiencies:

$$\begin{aligned} \eta_1 &= \frac{\Delta T x(x-1)}{(\frac{1}{2} + x)(x-1)T_h + \frac{1}{2}(x-1)T_c + (x^2-1)(1+x)/Z} \\ &= \frac{\Delta T \sqrt{1+Z\overline{T}}(\sqrt{1+Z\overline{T}}-1)}{(1+Z\overline{T})T_h + \sqrt{1+Z\overline{T}}T_c} = \eta_c \cdot \frac{\sqrt{1+Z\overline{T}}-1}{\sqrt{1+Z\overline{T}} + T_c/T_h} \\ \Rightarrow \quad \eta_2 &= \frac{T_h - T_c}{T_h - T_{am}} \cdot \frac{\sqrt{1+Z\overline{T}}-1}{\sqrt{1+Z\overline{T}} + T_c/T_h}. \end{aligned} \tag{5.36}$$

With state of the art technology the second-law efficiency never exceeds $\eta_2 \approx 0.2$. Furthermore, the best materials are expensive in their production and require the use of rare elements such as tellurium. The main application of thermoelectric elements is therefore in cooling devices, so-called **Peltier coolers**, which exploit the inverse thermoelectric effect, the **Peltier effect.** Here, an applied voltage and corresponding current lead to a heat flow across the thermoelectric device. Peltier coolers do not have moving parts and can be built in a very compact way, which makes them superior to other types of coolers for many applications. Another niche application for thermoelectrics is the energy supply for deep space missions where solar cells are no longer feasible due to the great distance to the sun. In these space probes a nuclear reaction is used as a heat source and a thermoelectric generator as a very lightweight heat engine free of moving parts.

6 Chemical exergy

In this chapter the exergy that a given amount of fuel in a given state can deliver is considered. Therefore, fundamental principles of the thermodynamics of chemical reactions are reviewed. As an example reaction, mostly the combustion of methane is used:

$$CH_4 + 2O_2 \rightarrow CO_2 + 2H_2O \tag{6.1}$$

6.1 Basic concepts

For the considerations in this chapter it is useful to first define standard conditions. The standard state of a pure substance at a specified temperature is the state – gaseous, liquid, or solid – in which it naturally exists at a pressure of $p^0 = p_{am} = 1$ bar. Note that the temperature is not part of the definition of standard conditions. For the calculation of standard reaction energies, such as reaction enthalpy or Gibbs free energy, the reactants and products of the reaction are assumed to be in their standard states at a particular temperature. This implies that the reactants are initially not mixed and that also the reaction products are separated into the individual components. These hypothetical initial and final states are depicted in Figure 6.1 for the example reaction equation (6.1).

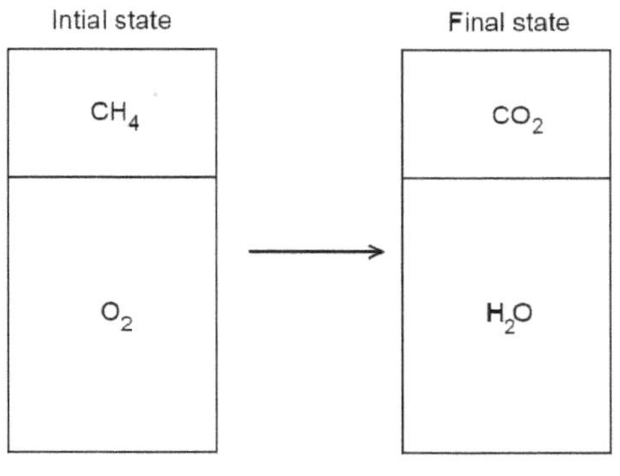

Fig. 6.1: Schematic of the initial and the final standard state of a chemical reaction. All four compartments shown are separated and filled with pure substances in their respective standard states, i.e. at a pressure of $p = 1$ bar.

In the hypothetical initial state, the reaction volume is separated into parts filled with the two reactant gases, methane and oxygen, in their stochiometric ratio of 1 : 2, both at standard pressure and at a specified temperature. Correspondingly, in the final state, the reaction products CO_2 and H_2O are also separated and in their standard state with a stochiometric ratio of 1 : 2. If the reaction takes place at a constant pressure p, the first law of thermodynamics equation (2.1) yields for the heat exchanged during the reaction

$$\begin{aligned} \Delta U &= U_2 - U_1 = Q + W = Q - p(V_2 - V_1) \\ \Rightarrow \quad Q &= (U_2 - U_1) + p(V_2 - V_1) = H_2 - H_1 =: \Delta H_r, \end{aligned} \tag{6.2}$$

where the heat exchanged during the reaction is called the **reaction enthalpy** ΔH_r. If $\Delta H_r > 0$ holds, the reaction is called endothermic, and heat has to be provided to the system, and if $\Delta H_r < 0$, the reaction is called exothermic. The sign of ΔH_r thus again follows the sign convention used throughout this book. If at a specified temperature all reactants and reaction products are in their standard states, i.e. all are at $p = 1$ bar and with an activity of $a = 1$ (see Section 6.2.1 below), the reaction enthalpy becomes the standard reaction enthalpy $\Delta H_r \to \Delta H_r^0$. For the methane combustion, the standard reaction enthalpy is identical to the upper heating value if the reaction proceeds at temperatures $0\,^\circ\mathrm{C} < T < 100\,^\circ\mathrm{C}$, with liquid water as part of the products. The heating value differs from the standard reaction enthalpy in this temperature region by the latent heat released through the condensation of water, as it is defined with water in the gaseous phase. As already discussed in Section 1.1, the latter value is of greater importance in the context of statistics of energy consumption as it is identical to the primary energy stored in one mole of fuel. An important material specific quantity for any chemical compound, which can be used to calculate the standard reaction enthalpy, is the **molar standard enthalpy of formation** Δh_f^0. For standard pressure $p = p_{am}$ and temperature $T = T_{am}$ the values for Δh_f^0 are tabulated for many different molecules. As a source for numerical values, in this book the *CRC Handbook of Chemistry and Physics* is used (David R. Lide, ed., *CRC Handbook of Chemistry and Physics, Internet Version 2005*, http://www.hbcpnetbase.com, CRC Press, Boca Raton, FL, 2005). The values of the enthalpies of formation correspond to the standard reaction enthalpy for the formation reaction of the compound from its constituent elements under standard conditions, as for example

$$\begin{aligned} C + O_2 &\to CO_2 \\ C + 2H_2 &\to CH_4 \end{aligned} \tag{6.3}$$

The reactants of these formation reactions, which are always made up of only one element, e.g. O_2, N_2, or C, do not need to be formed and consequently have an enthalpy of formation defined as $\Delta h_f^0 \equiv 0$. Conversely, the standard reaction enthalpy can be calculated by subtracting the sum of the standard enthalpies of formation for the reactants from the sum of the standard enthalpies of formation for the reaction products:

$$\Delta h_r^0 = \Sigma_{\text{prod.}} |\nu_j| \Delta h_f^0 - \Sigma_{\text{reac.}} |\nu_k| \Delta h_f^0 =: \Sigma_i \nu_i \Delta h_f^0. \tag{6.4}$$

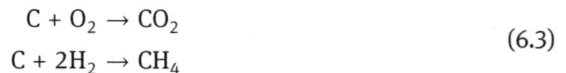

The ν_i are integers and are the **stoichiometric coefficients** of the different compounds involved. They are positive for reaction products and negative for reactants. In the methane combustion reaction their values are $\nu_{CH_4} = -1$, $\nu_{O_2} = -2$, $\nu_{CO_2} = 1$, and $\nu_{H_2O} = 2$.

If the reaction proceeds isobarically at $p = p_{am}$ and in an isothermal way at a temperature T, the mole-specific standard Gibbs free energy of the reaction Δg_r^0 can be calculated from the standard reaction enthalpies and the standard reaction entropies Δs_r^0. The latter are the entropy changes upon changing from a configuration of pure separated reactants in stoichiometric ratio in their standard states to a configuration of pure separated products in their standard states:

$$\Delta g_r^0 = \Delta h_r^0 - T \cdot \Delta s_r^0. \tag{6.5}$$

The change in the standard molar entropy during a reaction Δs_r^0 is the difference between the sum of the standard molar entropies of the products and reactants s_i^0 involved in the reaction. For the calculation of the standard entropies the third law of thermodynamics is employed, i.e. $s(T = 0\ \mathrm{K}) = 0\ \mathrm{J/(mol\ K)}$. The change in entropy of the pure substance when reversibly heating it from $T = 0$ K to the reaction temperature then has to be determined. A variety of phenomena such as phase transitions typically occur between $T = 0$ and the desired reaction temperature. For this reason only tabulated values are used within this book.

6.1.1 Example calculation

As an example, the standard reaction enthalpy of the methane combustion reaction equation (6.1) at ambient temperature can be calculated according to equation (6.4), using the enthalpies of formation of the reactants and products: $\Delta h_f^0(CH_{4[g]}) = -74.81$ kJ/mol, $\Delta h_f^0(CO_{2[g]}) = -393.51$ kJ/mol, and $\Delta h_f^0(H_2O_{[g]}) = -285.8$ kJ/mol:

$$\Delta h_r^0 = \Delta h_f^0(CO_{2[g]}) + 2 \cdot \Delta h_f^0(H_2O_{[g]}) - \Delta h_f^0(CH_{4[g]}) = -846.3\ \mathrm{kJ/mol}. \tag{6.6}$$

The reactant oxygen has a standard enthalpy of formation of zero, as it is the standard form in which oxygen atoms are found under standard conditions. The standard reaction enthalpy of the methane combustion is negative, i.e. the reaction is exothermal.

The standard molar Gibbs free energy Δg_r^0 of the methane combustion can be calculated with equation (6.5) using the tabulated standard entropies of the various compounds at ambient temperature: $s^0(CH_{4[g]}) = 186.2\ \mathrm{J/mol \cdot K}$, $s^0(O_{2[g]}) =$

205.0 J/mol · K, $s^0(CO_{2[g]}) = 213.6$ J/mol · K and $s^0(H_2O_{[l]}) = 69.6$ J/mol · K.

$$\begin{aligned}\Delta g_r^0 &= \Delta h_r^0 - T_{am} \cdot \Delta s_r^0 \\ &= \Delta h_r^0 - T_{am} \cdot (s^0(CO_{2[g]}) + 2s^0(H_2O_{[g]}) - s^0(CH_{4[g]}) - 2s^0(O_{2[g]})) \\ &= -774.8\ \text{kJ/mol}.\end{aligned} \tag{6.7}$$

The sign of the standard Gibbs free energy is also negative, i.e. it is an exergonic reaction and the system can perform useful work during the reaction.

6.2 The driving force of a chemical reaction

In a chemical reaction, the reactant compounds are continuously transformed into the product compounds. This means that reactant molecules are continuously taken out of the molecule ensemble at their chemical potential $\mu_{m,i}$, while product molecules are continuously added at their chemical potential $\mu_{m,j}$. The index 'm' at the chemical potential denotes a mole specific value measured in units of J/mol. In a system at constant pressure and temperature the Gibbs free energy is a function of the amounts of substances of the individual species only:

$$dG = \sum_i \mu_{m,i} dn_i. \tag{6.8}$$

As the amounts of substances n_k of all compounds are linked through the stoichiometry of the reaction, one parameter that links a small change in the amounts of substance of one arbitrary compound involved in the reaction dn_i to its stoichiometric coefficient ν_i is sufficient to describe all changes in the various compounds. It is called the **extent of reaction** ξ:

$$d\xi := \frac{dn_i}{\nu_i} = \frac{dn_j}{\nu_j}. \tag{6.9}$$

The extent of reaction ξ has the unit 'mole' and always lies between 0 and 1 mol, where the former limit is associated with the state of pure reactants and the latter one with the state of pure reaction products. Then, the change in the mole specific Gibbs free energy associated with a change in the extent of reaction is given by

$$\left(\frac{\partial G}{\partial \xi}\right)_{p,T} = \Sigma_k \nu_k \mu_{m,k}. \tag{6.10}$$

At constant pressure and temperature the change in the Gibbs free energy with the extent of reaction is the driving force for the reaction, as the system is always driven towards a minimum of G. Therefore, a nonzero value of $(\partial G/\partial \xi)_{p,T}$ leads to a change in the molecular composition of the system. More precisely, the conversion of reactants into products only proceeds as long as

$$\left(\frac{\partial G}{\partial \xi}\right)_{p,T} < 0. \tag{6.11}$$

The negative absolute value of this driving force of a chemical reaction is also referred to as **affinity**. In case of a positive sign of $(\partial G/\partial \xi)_{p,T}$, the reaction proceeds in the opposite direction, and in equilibrium the chemical reaction does not proceed, i.e. $(\partial G/\partial \xi)_{p,T} = 0$. G is then indeed minimized. Figure 6.2 depicts a typical course of G as a function of ξ for standard conditions.

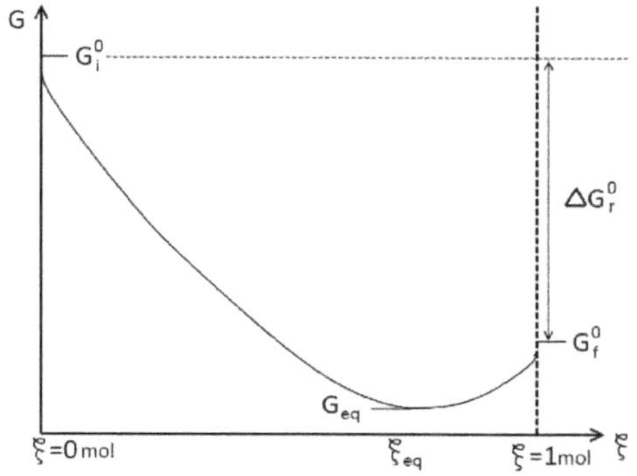

Fig. 6.2: Gibbs free energy G as a function of the extent of reaction ξ at $p = p_{am}$. The values G_0 and G_1 correspond to pure reactants and pure reaction products in their standard states, respectively. The equilibrium extent of the reaction is given by the minimum of $G(\xi)$.

G_i^0 denotes the sum of the Gibbs free energies of formation of the reactants in their standard states and G_f^0, correspondingly, the sum of the standard Gibbs free energies of formation of the reaction products. The difference between both is thus the molar standard Gibbs free energy of reaction ΔG_r^0.

6.2.1 Chemical activity

The definition of the standard reaction enthalpy and the Gibbs free energy states that all reactants and all reaction products are separated and in their standard state. However, for the reaction to occur, normally a slightly different system, where the components are each filling up the reaction volume, has to be assumed. This means that the entropy associated with the expansion of the reactants and products into the entire reaction volume has to be considered. For the further considerations it is useful to first consider an ideal gas which expands reversibly from an initial volume V_i to a final volume $V_f > V_i$. When the temperature T and thus the internal energy U are kept constant, a change in the entropy upon bringing the ideal gas from an initial state 'i' to a final state 'f' can be written as follows (see Section A.1 of the appendix):

$$0 = dU = TdS - pdV + \mu_\mathrm{m} dn \overset{dn=0}{\Rightarrow} dS = \frac{p}{T} dV = \frac{nR}{V} dV$$
$$\Rightarrow \quad \Delta S = \int_\mathrm{i}^\mathrm{f} dS = nR \cdot \ln\left(\frac{V_\mathrm{f}}{V_\mathrm{i}}\right) = -nR \cdot \ln\left(\frac{p_\mathrm{f}}{p_\mathrm{i}}\right). \tag{6.12}$$

To describe the situation relevant in the context of the initial state of a chemical reaction, several different, gaseous components 'k' are considered, each initially occupying a volume $V_{\mathrm{k,i}}$ in their standard state. These components then expand separately to an identical final volume, the reaction volume $V_{\mathrm{k,f}} = \sum_\mathrm{k} V_{\mathrm{k,i}}$. As all gaseous components were initially in their standard states, i.e. at $p = p^0$, the total pressure does not change during the expansion process. This situation is depicted for two species in Figure 6.3.

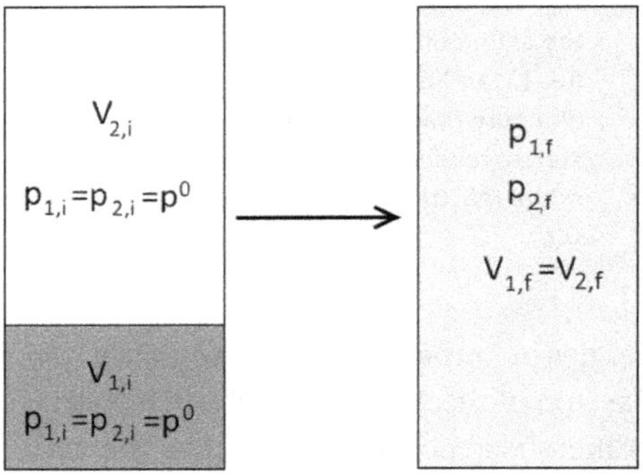

Fig. 6.3: Expansion of a gases '1' and '2' from compartments of the reaction volume into the full reaction volume.

Since the gases are assumed to be ideal, they do not interact, and the total mole specific entropy change going from the initial to the final state is called **entropy of mixture.** It is given by

$$\Delta s = \sum_\mathrm{k} \Delta s_\mathrm{k} = -R \cdot \sum_\mathrm{k} \ln\left(\frac{p_{\mathrm{k,f}}}{p_{\mathrm{k,i}}}\right) = -R \cdot \ln\left(\prod_k \frac{p_{\mathrm{k,f}}}{p_{\mathrm{k,i}}}\right). \tag{6.13}$$

In addition to this, in a real gas the interaction between the particles is also affected by the state change, which can be captured by multiplying the coefficient of fugacity $f_\mathrm{k}(p)$ to the pressure p_k as a correction factor. In this way the structure of the formulas for ideal gases can be kept. Per definition, $f_\mathrm{k} = 1$ holds for ideal gases. The change in entropy of species 'k' can then be written in terms of the chemical activity of species 'k':

$$\Delta s_\mathrm{k} = -R \cdot \ln(a_\mathrm{k}), \tag{6.14}$$

where the activity a_k of a gas is defined as the ratio between the partial pressure of species 'k' and the total pressure, corrected with the coefficient of fugacity:

$$a_k := f_k \cdot \frac{p_k}{p_{am}}. \tag{6.15}$$

The energy per mole required to insert species 'k' to the reaction volume at a constant temperature $\mu_{m,k}$ is lowered, compared to the standard state due to this entropy of mixture. This leads to the following expression:

$$\mu_{m,k} = \mu^0_{m,k} - T\Delta s_k = \mu^0_{m,k} + RT \cdot \ln(a_k), \tag{6.16}$$

where $\mu^0_{m,k}$ corresponds to the mole specific energy required to insert the component 'k' into a volume filled with component 'k' in its standard state, i.e. at $p_k = p^0$.

The derivation of equation (6.16) was only made for gases. However, the leading idea, namely the change of the entropy of the system upon diluting a species in the reaction volume, as shown in Figure 6.3, can also be applied to liquid solutions. The main difference is the definition of the chemical activity of the species involved. For stable liquids it is defined, in analogy to the case of ideal gases, as the molar fraction x_k of compound k in the entire solution multiplied by a correcting activity coefficient γ_k, accounting for the change in interaction of the species with the environment:

$$a_k := \gamma_k x_k = \gamma_k \frac{n_k}{\Sigma_i n_i}. \tag{6.17}$$

A different definition is used for the activity of solvated ions, e.g. Na^+, H^+, etc. Here, the concentrations of the species in a solvent such as water have to be considered. The activity of a dissolved species 'k' is then defined in the following way:

$$a_k := \gamma_k \frac{c_k}{c_0}, \tag{6.18}$$

with $c_0 = 1\ \mathrm{M} = 1\ \mathrm{mol/l}$ and the activity coefficient γ_k. Note that often the product $\gamma_k c_k$ is defined as activity. This definition has advantages when referring to chemical reaction rates, but then the activity is not dimensionless and is thus less convenient in the context of thermodynamics.

Solids are not filling up the reaction volume, leading to no entropy of mixture and, consequently, their activity is always unity: $a = 1$.

6.2.2 The driving force of a chemical reaction at a given state

If the reactants and reaction products are in a state different from their standard state, i.e. the activities of the chemical species are not all equal to one, the total change in the molar Gibbs free energy upon reaction Δg_r differs from the molar standard reaction Gibbs free energy Δg_r^0. The latter term can be stated in terms of the chemical potentials:

$$\Delta g_r^0 = \sum_k \nu_k \mu_{m,k}^0. \tag{6.19}$$

According to equation (6.16) the molar Gibbs free energy of reaction Δg_r when converting a set of mixed reactants to the set of mixed reaction products is thus given by

$$\begin{aligned}\Delta g_r &= \sum_k \nu_k \mu_{m,k} = \sum_k \nu_k(\mu_{m,k}^0 + RT\ln(a_k)) = \Delta g_r^0 + RT \cdot \sum_k \nu_k \ln(a_k) \\ &= \Delta g_r^0 + RT\ln\left(\prod_k a_k^{\nu_k}\right) = \Delta g_r^0 + RT \cdot \ln\left(\frac{\prod_{prod,j} a_j^{|\nu_j|}}{\prod_{reac,i} a_i^{|\nu_i|}}\right),\end{aligned} \tag{6.20}$$

where in the last term the product in the numerator is over the reaction products and in the denominator over the reactants, as the stoichiometric coefficients of the reaction products are positive and those of the reactants negative. In the course of the reaction, the activities of the species depend on the extent of reaction ξ, and equation (6.20) takes the form

$$\left(\frac{\partial G}{\partial \xi}\right)_{p,T} = \Delta g_r^0 + RT \cdot \ln\left(\frac{\prod_{prod,j} a_j(\xi)^{|\nu_j|}}{\prod_{reac,i} a_i(\xi)^{|\nu_i|}}\right) \tag{6.21}$$

6.3 The exergy of fuels

To determine the exergy of a chemical fuel, one has to calculate the maximal useful work extractable upon bringing the atoms contained in the fuel from their initial state, i.e. their chemical binding in the fuel, to their lowest energy natural form in the ambient state. In general, a chemical reaction is required to achieve this. If there are several chemical reactions possible, the one with the highest extractable work is chosen. For carbon-based fuels this means that one end product is always CO_2, and in the case of hydrocarbons water is also formed. Note that for a meaningful discussion of the exergy of a chemical fuel not only the fuel itself but also the other reactants taking part in the chemical reaction have to be included in the considerations. Thus, the term "exergy of a chemical fuel" actually means the "exergy of a set of reactants containing the fuel in question".

According to equation (2.11) the molar exergy of a closed system in state 'i', as for example the set of reactants, is given by

$$\Phi_{m,i} = (u_i - u_{am}) + p_{am}(v_i - v_{am}) - T_{am}(s_i - s_{am}), \tag{6.22}$$

where u, v, and s are mole specific, and in the ambient state the reaction products are also in chemical equilibrium with the surroundings, i.e. at their natural ambient activities. For the calculation of $\Phi_{m,i}$ it is useful to split the transition from the initial state to the ambient state into two steps. First, the burning of the fuel to the final state consisting of the mixed reaction products characterized by u_f, v_f, s_f and $a_{j,f}$ is considered. In the second step, the diffusion of these products in the atmosphere, i.e. the establishment of a chemical equilibrium between reaction products and the surrounding atmosphere, is taken into account. Furthermore, as the fuel is in general at ambient temperature and pressure, the reaction is carried out at $p = p_{am}$ and $T = T_{am}$. Both processes are illustrated in Figure 6.4 for the combustion of methane in air.

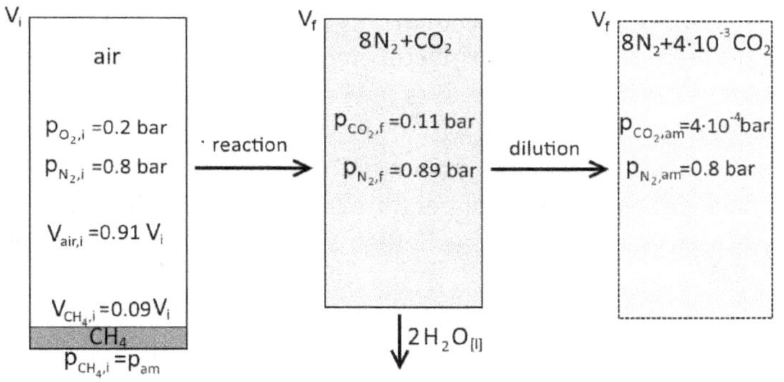

Fig. 6.4: Methane combustion in air at ambient pressure and temperature as a two-step process.

In the initial state 'i' the reactants have given activities $a_{k,i}$ and are in the correct stoichiometric ratio. The end product of the reaction is the mixture of reaction products. For the exergy calculation it is assumed that all reactants are fully converted into reaction products, i.e. that the equilibrium is only reached at $\xi \approx 1$ mol. The ambient state is characterized by the reaction products in natural concentrations in the atmosphere, which means that in a second step the product mixture diffuses into the atmosphere, thereby increasing its entropy. According to the definition of the exergy in equation (1.4), and taking into account that the internal energy is not affected by the diffusion of the products into the atmosphere, the molar exergy Φ_m can be written as

$$\begin{aligned}\Phi_m &= (u_i - u_f) + p_{am}(v_i - v_f) + p_{am}(v_f - v_{am}) - T_{am}(s_i - s_f) - T_{am}(s_f - s_{am}) \\ &= -\Delta h_r + T_{am}\Delta s_r - T_{am}(s_f - s_{am}) = -\Delta g_r - T_{am}(s_f - s_{am}) \\ &= \underbrace{-\Delta g_r^0 - RT_{am} \cdot \ln\left(\frac{\Pi_{prod,j} a_{j,f}^{\nu_j}}{\Pi_{reac,k} a_{k,i}^{\nu_k}}\right)}_{-\Delta g_r(T_{am})} \underbrace{-RT_{am} \cdot \ln\left(\Pi_{prod}\left(\frac{a_{j,am}}{a_{j,f}}\right)^{\nu_j}\right)}_{\text{diffusion in atmosphere}},\end{aligned} \tag{6.23}$$

where $a_{j,f}$ stands for the activities of the reaction products immediately after the reaction and $a_{j,am}$ for the corresponding activities of the product species in the ambient state. The variables u, v, s, h, and g are all mole specific. This means that the value for the exergy of a mole of a given fuel, or more exactly the exergy of a given set of reactants containing this fuel, can be calculated from tabulated values for the enthalpies of formation for the different products and reactants and their respective standard entropies and from the concentrations of the reactants in the initial and the products in the atmospheric state. The reaction part is by far larger than the second part, which accounts for the diffusion of the products into the atmosphere. Since the entropy increases in the second step, this term is positive and increases the molar exergy. However, this portion cannot be converted into useful work with traditional methods. Hence, as a rule of thumb, the maximal work extractable from a fuel is approximately given by the Gibbs free energy of reaction at ambient temperature $\Delta g_r(T_{am})$. A graphical representation of the exergy of a given set of reactants in their standard states is shown in Figure 6.5.

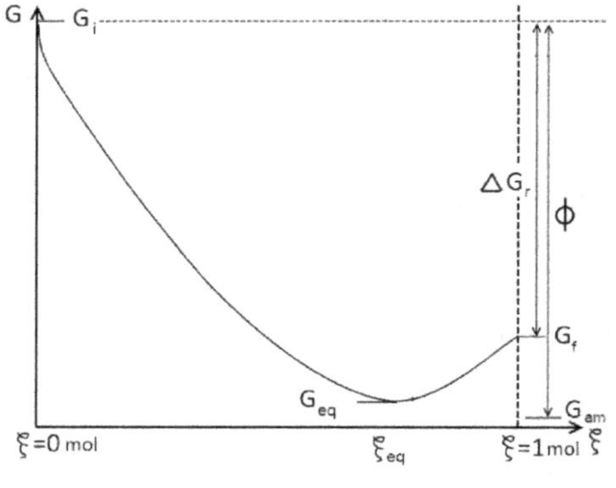

Fig. 6.5: Gibbs free energy G as a function of the extent of a chemical reaction ξ together with the exergy Φ of the fuels in state 'i'.

As an example, the methane combustion reaction (equation (6.1)) is considered, and an answer to the question which amount of exergy is stored in the set of reactants methane and oxygen is given using equation (6.23). In the ambient state the product CO_2 has a partial pressure of $p_{CO_2} = 4 \cdot 10^{-4} p_0$, and the reactant oxygen has a partial pressure of $p_{O_2} = 0.2$ bar. With these values and the value for Δg_r^0 given in equation (6.7), the following exergy is found:

$$\Phi_{m,CH_4+air} = -\Delta g_r^0 - RT_{am} \cdot \ln\left(\frac{a_{CO_2}}{a_{O_2}^2}\right) = 786.2\ \text{kJ/mol}. \tag{6.24}$$

6.4 Efficiency of the combustion process

The first-law efficiency of the combustion process is always $\eta_1 = 1$, as both the energy input and energy output are the reaction enthalpy Δh_r under the given conditions. For the second-law efficiency the temperature at which the heat is rejected by the combustion process has to be considered. The easiest case is the isothermal combustion process at temperature $T = T_h$. Here, the second-law efficiency is given by

$$\eta_2 = \frac{\Phi_{out}}{\Phi_{in}} = \frac{|\Delta h_r|\,(1 - T_{am}/T_h)}{\Phi_{m,CH_4+air}} = \frac{|\Delta h_r|}{|\Delta g_r|}\left(1 - \frac{T_{am}}{T_h}\right). \tag{6.25}$$

The temperature dependence of Δh_r is given by the heat capacities of reactants and reaction products via

$$\Delta h_r(T) = \Delta h_r(T_{am}) + \int_{T_{am}}^{T} \sum_i \nu_i c_p dT. \tag{6.26}$$

In the case of the methane combustion reaction in the temperature region considered, the integral in equation (6.26) is negligibly small, and $\Delta h_r(T) \approx \Delta h_r(T_{am})$ holds. This leads to a second-law efficiency for the isothermal methane combustion of

$$\eta_2 \approx \frac{\Delta h_r(T_{am})}{\Delta g_r(T_{am})}\left(1 - \frac{T_{am}}{T_h}\right). \tag{6.27}$$

In a more realistic process, for example in a gas turbine power plant, the combustion process is not isothermal but isobaric, and the system is an open one. In an open system in a steady flow equilibrium, the molar exergy of the set of reactants is given according to equation (3.15), which directly leads to equation (6.23) with the replacement $\Phi_m \rightarrow \psi_m$. The molar exergy of the open system is thus identical to the molar exergy of a closed system for all sets of reactants. It is now possible to adjust the mass flow and the fuel-to-air ratio in a way that one mole of reactants heats the flue gas from $T = T_c$ to $T = T_h$. Under the assumption of a nearly constant Δh_r and a constant average heat capacity $c_{p,av}$ for the air/reactant and residual air/product mixture, respectively, this leads to:

$$|\Delta h_r| = c_{p,av}(T_h - T_c) =: c_{p,av}\Delta T. \tag{6.28}$$

The exergy output $\psi_{m,out}$ and the second-law efficiency of the combustion process into the flue gas are then given by

$$\begin{aligned}\psi_{m,out} &= \int_{T_c}^{T_h} d\psi_m = \int_{T_c}^{T_h} c_{p,av}\left(1 - \frac{T_{am}}{T}\right) dT = \frac{|\Delta h_r|}{\Delta T}\int_{T_c}^{T_h}\left(1 - \frac{T_{am}}{T}\right) dT \\ &= |\Delta h_r|\left(1 - \frac{T_{am}}{\Delta T}\ln\left(\frac{T_h}{T_c}\right)\right) \\ \Rightarrow \quad \eta_2 &= \frac{\psi_{m,out}}{\psi_{m,in}} = \frac{\Delta h_r}{\Delta g_r^0}\left(1 - \frac{T_{am}}{\Delta T}\ln\left(\frac{T_h}{T_c}\right)\right).\end{aligned} \tag{6.29}$$

The second-law efficiency of the methane combustion for isothermal and isobaric processes as given in equations (6.27) and (6.29) is shown in Figure 6.6. For other fossil fuels, such as e.g. coal, the overall picture is similar. For the temperature values shown the reaction product water is gaseous.

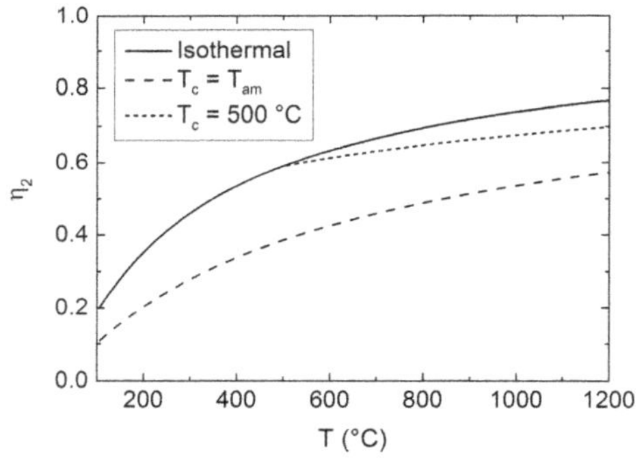

Fig. 6.6: Second-law efficiencies as a function of T_h for the methane combustion in an isothermal process (*solid curve*) and in isobaric processes with $T_c = T_{am}$ (*long-dashed curve*) and $T_c = 500\,°C$, which is a realistic lower temperature level for the combustion chamber in gas turbine power plants (*short-dashed curve*).

The combustion process is an irreversible process, as it takes place spontaneously at the temperatures shown. The exergy loss due to this irreversible process is rather substantial at low output temperatures. Only at high temperatures is a significant part of the value of the fuel retained. This loss of exergy can be addressed by using electrochemical devices which directly convert the chemical exergy of the fuels into electricity and will be discussed in the following chapter.

7 Electrochemical energy conversion

In this chapter the direct conversion of chemical into electrical energy is discussed. This means that instead of a combustion or more general heat release in an exothermic reaction, and a subsequent heat engine step, the combustion reaction of the chemical fuels runs in separate parts of an appropriate device, and the output is directly electrical energy. Such a device is an electrochemical energy converter, and the most well-known examples for such converters are batteries and fuel cells. The main advantage of the direct conversion route as opposed to the intermediate heat production is that an electrochemical device is not limited by the Carnot efficiency, as this restriction is only relevant for heat engines. Instead, the maximal external voltage $U^{\max}$ is expected to be related to the exergy stored in the fuels as electrical energy is pure exergy:

$$U^{\max} = -\frac{\Delta g_r^0}{nF}, \tag{7.1}$$

where n is the number of electrons transferred in the conversion of one fuel molecule. In the course of this chapter, the basic concepts of electrochemistry as well as its application to the most common electrochemical energy converters will be discussed. Moreover, it will be shown that for reactants in their standard state, equation (7.1) holds.

7.1 Electrochemistry

Electrochemistry deals with charge carrier transfer processes across an electrode|electrolyte interface. During such a charge transfer process the charge carrier has to be converted from a form which is mobile in the solid state electrode, namely an electronic state of the electrode like a free electron or a free hole, into a form which is mobile in the electrolyte, namely an ion. This conversion is always associated with a chemical redox reaction at the electrode|electrolyte interface. During this reaction the electrode either acts as a source or as a sink of electrons. In the former case, electrons are transferred to a species in the electrolyte, say species A, which is then reduced to A^-, and the electrode is called a **cathode**. Conversely, in the latter case electrons are absorbed by the electrode from a chemical species, say species B, which is thereby oxidized to B^+, and the electrode is called an **anode**. The currents associated with these respective processes are then called cathodic and anodic, respectively:

$$\begin{aligned} &A + e^- \rightarrow A^- \quad \text{cathodic reaction} \\ &B \rightarrow B^+ + e^- \quad \text{anodic reaction.} \end{aligned} \tag{7.2}$$

Every electrochemical reaction can in principle proceed in both directions, i.e. for example,

$$A + e^- \rightleftharpoons A^-. \tag{7.3}$$

This means that depending on the energetics of the electrode|electrolyte interface, the reactions just mentioned can also proceed in the opposite direction, and the description of an electrode as a cathode or anode is not an intrinsic property of the electrode but depends on the actual direction of the net current flow. The latter is determined not only by the electrode in question and the chemical environment into which it is immersed, but also by the second electrode in the electrolyte, which is necessary to close the electrical circuit. The energetics of the two interfaces, specifically the positions of the electrochemical potentials of the electrons with respect to the energy levels of the chemical species involved in the redox reactions, is thus central to electrochemical considerations. It will be discussed in the following sections.

7.1.1 The standard hydrogen scale

To discuss the energetics of an electrode|electrolyte interface, it is first necessary to introduce an energetic zero point against which the electrochemical potentials of the electrons in both material phases are measured. For these energetic considerations the electrode|electrolyte interface is always assumed to be in equilibrium such that the oxidation and reduction reaction always proceed equally fast and no net current flows. The natural energy reference level for the charge carriers in the electrode, i.e. outside the electrolyte, is the vacuum level which was also used in this book so far. This choice is, however, very inconvenient for dissolved species, as the energetics of the electron entering or leaving the electrolyte then has to be taken into account, which is a tedious process theoretically and extremely difficult to measure in experiments. To elegantly get around these problems, a reference reaction in the electrolyte is instead introduced as the zero electron energy reference level. This reaction is the hydrogen reduction/oxidation reaction

$$\frac{1}{2}H_2 \rightleftharpoons H^+ + e^-, \tag{7.4}$$

where the reactants are given under standard conditions ($pH = 0$, that is, $a_{H^+} = 1$, $p(H_2) = 1$ bar, $T = T_{am}$) in an aqueous electrolyte. For energy considerations, an electrochemical reaction at an electrode is then modeled in the following way:

$$A_{[aq]} + e^- \rightleftharpoons A^-_{[aq]} \quad \Rightarrow \quad A_{[aq]} + \frac{1}{2}H_{2[aq]} \rightleftharpoons A^-_{[aq]} + H^+_{[aq]}. \tag{7.5}$$

Translating this to an actual electrochemical setup involving an electrode immersed in an electrolyte, the reaction given in equation (7.5) is separated into two half reactions in two different compartments, as shown in Figure 7.1.

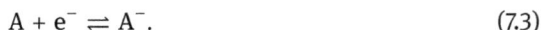

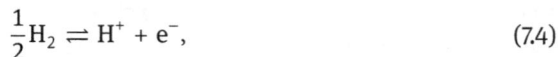

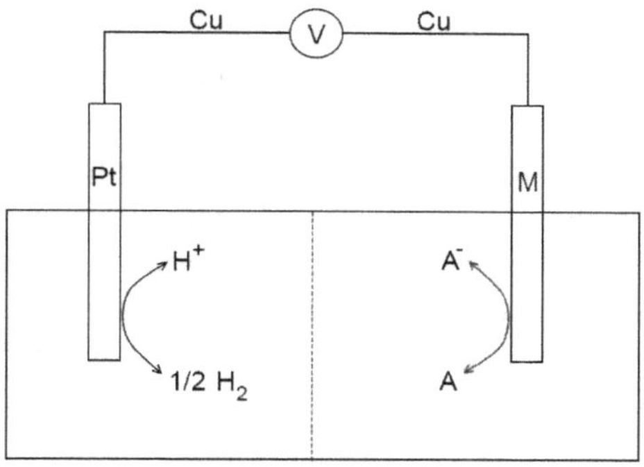

Fig. 7.1: Schematic of an electrochemical setup with a reference reaction.

In such a configuration a test charge participating in the reaction $1/2H_2 \rightleftharpoons H^+$ has to take the detour around the external circuit to participate in the half reaction $A \rightleftharpoons A^-$. The test charge then originates from the electrolyte and is rejected again into the electrolyte, and thus the energetics on its entering or leaving the electrolyte cancel out. It is very important to note that while the reactants cannot mix in the two chambers, the electrostatic potential of the bulk electrolyte is identical in both compartments, as the separating membrane is chosen to be electrically conductive. An electrode|electrolyte setting as in the left hand side of Figure 7.1 is called a **standard hydrogen electrode (SHE)** for $p = p^0$ and $a_{H^+} = 1$, and the electrostatic potential of the electrons in the platinum electrode is defined as $\varphi_{Pt} \equiv 0$. The hydrogen scale with this SHE potential as reference point is always used in electrochemistry when aqueous electrolytes are used. As the conditions in the standard hydrogen electrode itself are fixed, the voltage U between the two copper leads is only dependent on the redox couple A/A^- of the right-hand compartment. This voltage is called the **electrode potential** E_{el}, or **redox potential,** the term potential being used rather than voltage, as it is only dependent on one electrode. When more than one redox couple is present in the right-hand compartment, the voltage measured against SHE depends on all of these couples. This voltage is called **open circuit potential** and is subsequently denoted as E_{el}^{oc}.

7.1.2 Origin of the electrode potential

As already laid out in Section 5.1.1, changes in the electrostatic potential only occur at interfaces due to the formation of space charge layers. In equilibrium the electrostatic potentials stay constant in bulk materials. This is also true for bulk electrolytes. The potential drop across the membrane also vanishes as long as all species that can cross it have identical activities in both compartments. At the electrode|electrolyte interfaces a space charge layer, as shown in Figure 5.2, forms which is responsible for an electrostatic potential drop across the interfaces. In electrochemistry this space charge layer is called **double layer.** The part of the double layer at an electrode|electrolyte interface extending into the electrolyte can be thought of as consisting of two parts. In the (outer) Helmholtz layer, extending one diameter of an ion solvation shell from the electrode surface into the electrolyte, the potential drops linearly. The region further away from the electrode is called the Gouy–Chapman layer and is characterized by an exponential drop of the electrostatic potential from the value at the outer boundary of the Helmholtz layer to the bulk value. The double layer as a whole acts as a capacitor, and in the case of a negligible extension of the Gouy–Chapman layer it can be approximated as a plate capacitor with a plate separation of the typical length scale of about 1 nm. For an aqueous electrolyte ($\varepsilon_{H_2O} \approx 8$ close to the surface) this leads to an area-specific capacitance in the order of

$$C_{dl}/A_{el} = \frac{\varepsilon_0 \cdot \varepsilon_{H_2O}}{d_{dl}} \approx 1 \cdot 10^{-5}\ \mathrm{F/cm^2}. \tag{7.6}$$

This value is very high compared to conventional electronic plate capacitors, making electrolyte capacitors an important application of electrochemistry. In contrast to conventional capacitors the capacitance is also voltage dependent, i.e. $C \to C(U)$ for electrolyte capacitors.

The changes in the electrostatic potential due to the space charge layers are shown in Figure 7.2 for a general electrochemical setup.

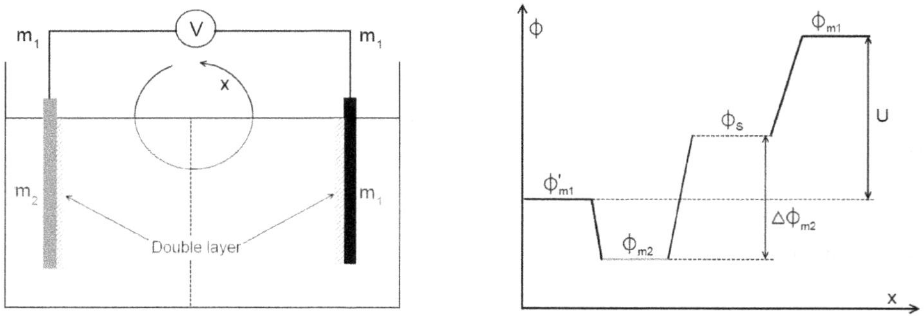

Fig. 7.2: Schematic of a voltage measurement in an electrochemical setup (*left*). Different electrical lead and electrode materials m_1 and m_2 are shown, as well as the double layers at the respective electrodes. An example for a potential profile along the x-direction (indicated in the left picture) is shown (*right*). The *vertical dashed line* in the left picture indicates a membrane which is electrically conductive while still maintaining different chemical environments on both sides. All potential drops occur across the interfaces.

The electrostatic potential drop $\Delta\varphi_{m2}$ across the double layer at the interface made from material m_2 in Figure 7.2 can be written as follows:

$$\Delta\varphi_{m2} = \varphi_{m2} - \varphi_s, \tag{7.7}$$

where φ_s is the electrostatic potential of the bulk electrolyte. The voltage U measured by the voltmeter is the sum over the three electrostatic potential drops occurring at the three different material transitions:

$$\begin{aligned} U &= (\varphi_{m1} - \varphi_s) + (\varphi_s - \varphi_{m2}) + (\varphi_{m2} - \varphi'_{m1}) =: \Delta\varphi_{m1} - \Delta\varphi_{m2} + (\varphi_{m2} - \varphi'_{m1}) \\ &= \Delta\varphi_{m1} - \Delta\varphi_{m2} + \frac{1}{F}\left(\mu_{m,m2} - \tilde{\mu}_{m,m2} - \left(\mu_{m,m1} - \tilde{\mu}'_{m,m1}\right)\right) \underbrace{=}_{\tilde{\mu}_{m,m2}=\tilde{\mu}'_{m,m}} \\ &= \underbrace{\Delta\varphi_{m1}}_{=:\Delta\varphi_{el,1}} - \underbrace{\left(\Delta\varphi_{m2} - \tfrac{1}{F}\left(\mu_{m,m2} - \mu_{m,m1}\right)\right)}_{=:\Delta\varphi_{el,2}}, \end{aligned} \tag{7.8}$$

where $\Delta\varphi_{el,1}$ and $\Delta\varphi_{el,2}$ are the differences in electrostatic potentials between the respective lead in the voltmeter and the electrolyte. They are not measurable themselves: only their sum is measurable. In equation (7.8) the fact was used that the electrochemical potential of the electrons is identical from the lead of the voltmeter to the redox couple in equilibrium in each compartment. Note that while the material of the electrode on the right-hand side in Figure 7.2 is in general not the material of the electronic leads, this difference is not important for the evaluation of $\Delta\varphi_{m2}$, as the value for $\Delta\varphi_{m2}$ stays the same whether or not there is an additional step in between. If the right-hand compartment in Figure 7.2 is an SHE, the expression $U = E_{el,2}$ holds, i.e. the voltage measured is equal to the electrode potential as defined above.

7.1.3 Electrode potential and cell voltage

In an electrochemical standard setup, as shown in Figure 7.1, a change in the free energy of the reactants of the reaction given in equation (7.5) upon an infinitesimal progress of reaction $(\partial G/\partial\xi)_{p,T}$ is associated with the transport of test charges around the circuit. These test charges cross both electrode|electrolyte interfaces and the voltmeter. When measuring against SHE, the overall setting and the two half-reactions, as shown in in Figure 7.1, are present. In the following derivation the changes in the electrostatic potential felt by an electron originating from an H_2 molecule in the left-hand compartment when traveling to the left-hand lead of the voltmeter, are compared to those felt by an electron originating from an A^- ion in the right-hand compartment when traveling to the right-hand lead of the voltmeter. Under equilibrium conditions, the electrochemical potentials of the electrons in the electrolyte and in the electrodes are identical. This means that the electrochemical potential of the electrons in the bulk electrolyte $\tilde{\mu}_{m,s}$ has to be defined, which is done in the following way for the two electrodes in question:

$$\begin{aligned} \tilde{\mu}^{left}_{m,s} &:= 1/2\mu_{m,H_2} - \mu_{m,H^+} - F\varphi_s \\ \tilde{\mu}^{right}_{m,s} &:= \mu_{m,A^-} - \mu_{m,A} - F\varphi_s, \end{aligned} \tag{7.9}$$

where φ_s denotes the electrostatic potential of the bulk electrolyte. As the electrochemical potential of the electrons is equal on both sides of each interface, the electrostatic potential drop across them can be linked to the difference in the chemical potentials of the electrons in both material phases, respectively:

$$
\begin{aligned}
&\tilde{\mu}_{m,s}^{left} = \tilde{\mu}_{m,Pt} = \mu_{m,Pt} - F\varphi_{Pt} \\
\Rightarrow \quad & \Delta\varphi_{Pt} = \varphi_{Pt} - \varphi_s = \frac{1}{F}\left(\mu_{m,Pt} - 1/2\mu_{m,H_2} + \mu_{m,H^+}\right) \\
&\tilde{\mu}_{m,s}^{right} = \tilde{\mu}_{m,M} = \mu_{m,M} - F\varphi_M \\
\Rightarrow \quad & \Delta\varphi_M = \varphi_M - \varphi_s = \frac{1}{F}\left(\mu_{m,M} - \mu_{m,A^-} + \mu_{m,A}\right).
\end{aligned}
\tag{7.10}
$$

Here, $\mu_{m,Pt,M}$ and $\varphi_{Pt,M}$ are the chemical and electrostatic potentials in the two electrodes, respectively. Going one step further, both electrons cross the last interface on their path to the respective leads of the voltmeter. At this metal|copper interface the electrochemical potentials of the electrons are again in equilibrium, which introduces another drop in the electrostatic potential:

$$
\begin{aligned}
\Delta\varphi_{left} &:= \varphi_{Cu,left} - \varphi_s = (\varphi_{Cu,left} - \varphi_{Pt}) + \Delta\varphi_{Pt} = \frac{1}{F}(\mu_{m,Cu} - \mu_{m,Pt}) + \Delta\varphi_{Pt} \\
\Delta\varphi_{right} &:= \varphi_{Cu,right} - \varphi_s = (\varphi_{Cu,right} - \varphi_M) + \Delta\varphi_M = \frac{1}{F}(\mu_{m,Cu} - \mu_{m,M}) + \Delta\varphi_M.
\end{aligned}
\tag{7.11}
$$

The electrode potential is then the voltage measured by the voltmeter, as the left-hand electrode is an SHE. Combining equations (7.9)–(7.11), the following term for the electrode potential of the right-hand electrode is found:

$$
\begin{aligned}
E_{el,right} &= \varphi_{Cu,right} - \varphi_{Cu,left} = \frac{1}{F}(\mu_{m,Pt} - \mu_{m,M}) + \Delta\varphi_M - \Delta\varphi_{Pt} \\
&= \frac{1}{F}(-\mu_{m,A^-} + \mu_{m,A} + 1/2\mu_{m,H_2} - \mu_{m,H^+}).
\end{aligned}
\tag{7.12}
$$

By replacing the hydrogen reaction $1/2H_2 \rightleftharpoons H^+ + e^-$ by the more general reaction $B^+ + e^- \rightleftharpoons B$, the present derivation can be used to calculate the equilibrium cell voltage U^{eq} of any electrochemical full reaction $A + B \rightarrow AB$:

$$
U^{eq} = E_{el,A/A^-} - E_{el,B/B^+} = \frac{1}{F}(\mu_{m,A} + \mu_{m,B} - \underbrace{(\mu_{m,A^-} + \mu_{m,B^+})}_{=\mu_{m,AB}}) = -\frac{\Delta g_r}{F} \tag{7.13}
$$

Equation (7.13) indeed states that the exergy stored in a set of fuels can theoretically be fully converted to electrical energy, as naively assumed at the outset of this chapter in equation (7.1).

7.1.4 The Nernst equation

In the previous section a link between the electrode potential when measrued against SHE and the chemical potential of the species involved was established. In turn, in Section 6.2.2 the chemical potential of a species was linked to its activity. Putting both together, the electrode potential of an electrode immersed in an electrolyte with the species A and A^- can then be calculated using the **Nernst equation**:

$$\begin{aligned} E_{el} &= -\frac{1}{F}\sum_{i} \nu_i \mu_{m,i} = \underbrace{-\frac{1}{F}\sum_{i} \nu_i \mu_i^0}_{=:E_{el}^0} - \frac{RT}{F}\ln\left(\prod_{i} a_i^{\nu_i}\right) \\ &= E_{el}^0 + \frac{RT}{F}\ln\left(\frac{a_A}{a_{A^-}}\right). \end{aligned} \tag{7.14}$$

The term E_{el}^0 stands for the standard electrode potential, i.e. the potential when all chemical species at the electrode are in their standard state. A more generalized formulation of the Nernst equation is as follows:

$$E_{el} = E_{el}^0 + \frac{RT}{nF}\ln\left(\frac{\prod_{ox} a_i^{|\nu_i|}}{\prod_{red} a_j^{|\nu_j|}}\right), \tag{7.15}$$

where all chemical species participating in the electrochemical reaction at a single electrode in their oxidized and reduced form are considered and n is the number of electrons involved in the chemical half reaction. The activities of the oxidized species are in the numerator, as an increased presence of these species close to the electrode leads to an easier pathway for the electrons to leave the electrode and consequently a more positive equilibrium electrode potential.

When two electrodes are immersed into an electrolyte forming an electrochemical cell, an overall chemical reaction consisting of the two half reactions at the electrodes occurs. Under equilibrium conditions, the cell voltage U^{eq} can be calculated as the difference between the two electrode potentials (measured both against SHE). The electrode potentials in turn are determined using the Nernst equation (equation (7.15)):

$$\begin{aligned}
U^{\mathrm{eq}} &= E_{\mathrm{el},1} - E_{\mathrm{el},2} \\
&= E^0_{\mathrm{el},1} + \frac{RT}{nF} \cdot \ln\left(\frac{\prod_{\mathrm{ox},1} a_{\mathrm{i}}^{|\nu_{\mathrm{i}}|}}{\prod_{\mathrm{red},1} a_{\mathrm{j}}^{|\nu_{\mathrm{j}}|}}\right) - \left(E^0_{\mathrm{el},2} + \frac{RT}{nF} \cdot \ln\left(\frac{\prod_{\mathrm{ox},2} a_{\mathrm{i}}^{|\nu_{\mathrm{i}}|}}{\prod_{\mathrm{red},2} a_{\mathrm{j}}^{|\nu_{\mathrm{j}}|}}\right)\right) \\
&= U^0 + \frac{RT}{nF}\left(\ln\left(\frac{\prod_{\mathrm{reac},1} a_{\mathrm{i}}^{|\nu_{\mathrm{i}}|}}{\prod_{\mathrm{prod},1} a_{\mathrm{i}}^{|\nu_{\mathrm{i}}|}}\right) - \ln\left(\frac{\prod_{\mathrm{prod},2} a_{\mathrm{i}}^{|\nu_{\mathrm{i}}|}}{\prod_{\mathrm{reac},2} a_{\mathrm{j}}^{|\nu_{\mathrm{j}}|}}\right)\right) \\
&= U^0 + \frac{RT}{nF}\ln\left(\frac{\prod_{\mathrm{reac}} a_{\mathrm{j}}^{|\nu_{\mathrm{j}}|}}{\prod_{\mathrm{prod}} a_{\mathrm{i}}^{|\nu_{\mathrm{i}}|}}\right) = U^0 + \frac{RT}{nF}\ln\left(\prod_{\mathrm{k}} a_{\mathrm{k}}^{-\nu_{\mathrm{k}}}\right).
\end{aligned} \tag{7.16}$$

The standard cell voltage U^0, where all species are in their respective standard states, is defined by the change in the Gibbs free energy of the full reaction in the standard state Δg^0_{r}:

$$U^0 = -\frac{1}{nF}\Delta g^0_{\mathrm{r}}. \tag{7.17}$$

In the case of the important hydrogen-combustion/water-splitting reaction

$$2\mathrm{H}_2 + \mathrm{O}_2 \rightleftharpoons 2\mathrm{H}_2\mathrm{O} \tag{7.18}$$

the standard cell voltage at $T = T_{\mathrm{am}}$ is given by

$$U^0 = -\frac{\Delta g^0_{\mathrm{r}}}{nF} = -\frac{1}{4F}\left(2\Delta h^0_{\mathrm{f}}(\mathrm{H_2O_{[l]}}) - T_{\mathrm{am}}\Delta s^0_{\mathrm{r}}\right) = 1.23\ \mathrm{V}. \tag{7.19}$$

7.1.5 Electrochemical voltage sources

To conclude the previous sections, the built-in asymmetry in electrochemical setups leading to the external voltage will now be discussed. The externally measured voltage corresponds to a change in the electrochemical potential of the electrons when taking the path indicated in Figure 7.2. The question is then why the electrochemical potential is actually different between the two leads of the voltmeter when in equilibrium it is unchanged at all the interfaces considered, i.e. where the asymmetry leading to the external voltage is built into the setup. The answer is that while the two compartments of the electrolyte are connected in an electrically conducting way, the membrane in the center prevents a mixing of the chemical species on both sides. Thus, the electrostatic potential stays constant across the membrane while the chemical environment changes. While in equilibrium with their respective chemical environments, both electrodes are then interacting with two different chemical environments. The measured voltage is thus caused by the fact that the electrons, which themselves do

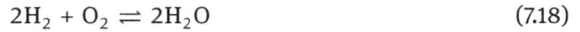

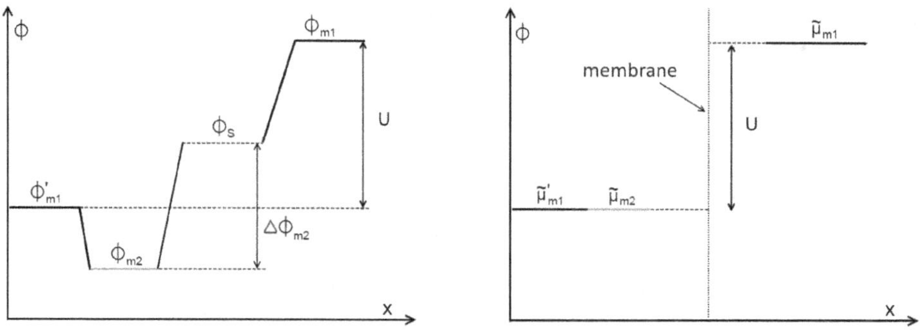

Fig. 7.3: Schematic comparison of the electrostatic potentials in an electrochemical system as shown in Figure 7.2 and the electrochemical potentials of the electrons in both leads. The latter jumps upon crossing the electrolyte, as both electrodes interact with different chemical environments, due to the electrically conducting membrane in the system.

not exist in the electrolyte, interact with different chemical environments separated by the membrane. This is illustrated in Figure 7.3, where the electrostatic potential taken from Figure 7.2 is contrasted with the electrochemical potential of the electrons in the solid state material phases.

The membrane, a part of the cell which might naively be regarded as nonessential because it does not take part in the electrochemistry of the electrodes, lies in this regard at the center of the function of the electrochemical device as a voltage source. In this picture the measured voltage is then linked to the change in the Gibbs free energy upon the full reaction Δg_r in an intuitive way. For each reacted molecule, n electrons have to take a detour around the circuit while at the same time transporting the full difference in free energy upon reaction. This energy difference is then measured in the voltmeter, directly leading to

$$\Delta g_r = -n(\tilde{\mu}_{\mathrm{m,an}} - \tilde{\mu}_{\mathrm{m,cat}}) = -nF(\varphi_{\mathrm{an}} - \varphi_{\mathrm{cat}}) = -nFU, \tag{7.20}$$

where $\tilde{\mu}_{\mathrm{an,cat}}$ and $\varphi_{\mathrm{an,cat}}$ are the molar electrochemical potentials of the electrons and the electrostatic potentials in anode and cathode, respectively. Equation (7.20) shows that a change in the free energy inside the electrolyte is identical to a transport of free energy out of the electrolyte via electrons in the electrodes. Apart from the diffusion of the reaction products into the atmosphere, the full exergy of the chemical fuels can thus be converted into electrical exergy in an electrochemical setup.

7.2 Electrochemical energy conversion

During direct electrochemical energy conversion, the energy stored in a set of reactants is directly converted into electrical energy by feeding the reactants to two electrodes in different compartments and thus splitting the reaction into two electrochemical half-reactions. In this way the irreversible combustion of the reactants is left out, and the corresponding loss of exergy can be circumvented. The electron transfer between both electrodes is directed through an electrical consumer, and the electric circuit is closed by the electrolyte. Direct electrochemical energy converters are called **galvanic cells.** A sketch of such a cell is shown in Figure 7.4 for the following hypothetical chemical reaction:

$$A + B \rightleftharpoons A^- + B^+ \rightleftharpoons AB. \tag{7.21}$$

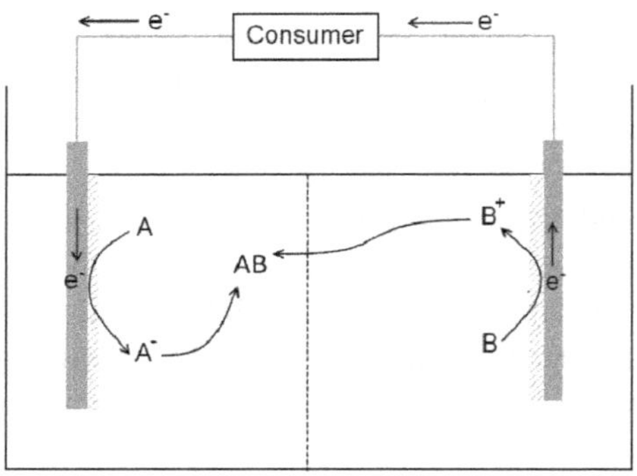

Fig. 7.4: Schematic of a galvanic cell with the overall reaction $A + B \rightarrow A^- + B^+ \rightarrow AB$.

Note that at least one chemical species, in Figure 7.4 the species B^+ always has to be able to cross the membrane in order to close the electrical circuit. This does not affect the purpose of the membrane to separate the chemical environments at the two electrodes, as long as the reactants of the full reaction, in Figure 7.4 A and B cannot cross it. In the ideal case, i.e. in the absence of additional reactions such as oxide formation, the open circuit cell voltage U_{oc} is then identical to the equilibrium cell voltage given in equation (7.16).

7.2.1 Maximal efficiency

For the direct electrochemical energy conversion realized in a galvanic cell, it is in principle possible to keep the reactants and products separated during the reaction. Neglecting the contribution of the diffusion of the products into the atmosphere the molar exergy of the set of reactants in their standard state Φ_m^0 can in principle be fully used:

$$\Phi_{\mathrm{m}}^{0} \approx \Delta g_{\mathrm{r}}^{0} = -nFU^{0}. \tag{7.22}$$

As the primary energy stored in one mole of fuels is defined as $\Delta h_{\mathrm{r}}^{0}$, a maximal possible first-law efficiency called **thermal efficiency** η_{th} is defined for galvanic cells in analogy to the Carnot efficiency for thermal engines:

$$\eta_{\mathrm{th}} \approx \frac{\Phi_{\mathrm{m}}}{\Delta h_{\mathrm{r}}^{0}} = \frac{\Delta g_{\mathrm{r}}^{0}}{\Delta h_{\mathrm{r}}^{0}} = 1 - T\frac{\Delta s_{\mathrm{r}}^{0}}{\Delta h_{\mathrm{r}}^{0}}. \tag{7.23}$$

Note that η_{th} can take values larger than one, as the sign of the standard reaction entropy $\Delta s_{\mathrm{r}}^{0}$ can take both signs, while $\Delta h_{\mathrm{r}}^{0} < 0$ always holds when energy is to be gained from the reaction. The temperature dependence of the first law efficiency at constant pressure is determined by the (negative) ratio of the reaction entropy and enthalpy:

$$\left(\frac{\partial \eta_{\mathrm{th}}}{\partial T}\right)_{\mathrm{p=p^0}} \approx -\frac{\Delta s_{\mathrm{r}}^{0}(T)}{\Delta h_{\mathrm{r}}^{0}(T)}. \tag{7.24}$$

A large change in the entropy caused by the reaction leads to a strong temperature dependence of the thermal efficiency η_{th}. The reaction entropy is dominated by the translational degrees of freedom of the reactants and products. It is significantly higher for gases than for other material phases, which makes it possible to estimate the temperature dependence of $\eta_{\mathrm{th}}$by just comparing the amounts of gaseous compounds of reactants and products Δn_{gas} as done for the following examples:

$$\begin{aligned}
&\mathrm{C + O_2 \rightarrow CO_2}: && \Delta n_{\mathrm{gas}} = 0\,\mathrm{mol} \;\Rightarrow\; \Delta s_{\mathrm{r}}^{0} \approx 0\\
&\mathrm{2C + O_2 \rightarrow 2CO}: && \Delta n_{\mathrm{gas}} = 1\,\mathrm{mol} \;\Rightarrow\; \Delta s_{\mathrm{r}}^{0} > 0\\
&\mathrm{2H_2 + O_2 \rightarrow 2H_2O_{[l]}}: && \Delta n_{\mathrm{gas}} = -3\,\mathrm{mol} \;\Rightarrow\; \Delta s_{\mathrm{r}}^{0} < 0.
\end{aligned} \tag{7.25}$$

The temperature effect on the thermal efficiency η_{th} can be very pronounced. Note that the standard reaction entropy $\Delta s_{\mathrm{r}}^{0}$ can be smaller than zero if the amount of gaseous products is smaller than the amount of gaseous reactants, as in the third example in equation (7.25). In this very important example, low temperature operations are favored over high temperature operations, as η_{th}decreases with temperature. It is instructive to compare the temperature dependence of the thermal efficiency of the electrochemical hydrogen combustion given in the last line of equation (7.25) with the temperature dependence of the maximal efficiency of a thermal engine driven with conventional hydrogen combustion as a heat source, i.e. with η_{C}. Such a comparison is shown in Figure 7.5.

Typically, a comparison between the thermal efficiency η_{th} and the Carnot efficiency η_{C}, i.e. the solid and the dashed curve in Figure 7.5, is seen as a straight-forward way of assessing which kind of process is preferred at a given temperature. In this case, above a temperature of 900 °C the combustion and subsequent heat engine pathway

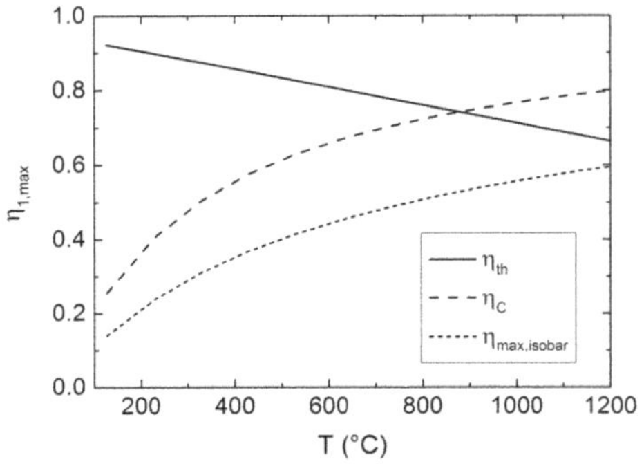

Fig. 7.5: Comparison between the thermal efficiency η_{th} for the water formation reaction (*solid curve*), the Carnot efficiency η_C (*long-dashed curve*) and the maximal efficiency for an isobaric combustion process (*short-dashed curve*).

seems to be preferable. This is, however, not quite the case, as the decrease of the thermal efficiency with temperature already factors in the energy cost for heating the fuel and reaction products to the desired temperature. This is not the case for the Carnot efficiency, where an isothermal combustion is assumed. A fair comparison thus takes the maximal efficiency for an isobaric heating process, as the reference curve for the heat engine process (*short-dashed line* in Figure 7.5) which shifts the crossing point between both curves to significantly higher temperatures. As evident from Figure 7.5) an electrochemical pathway is thus the superior choice for all temperatures shown. The same conclusion can also be drawn considering the second-law efficiencies. For an electrochemical energy converter its optimal value is given by

$$\eta_2 = \frac{\Delta g_r^0}{\Delta g_r^0} = 1, \tag{7.26}$$

while it is always smaller than one for heat engines, as the exergy content of the reaction heat is always smaller than the exergy content of the fuels themselves.

7.2.2 Efficiency of a realistic galvanic cell

So far the cell voltages were calculated for the equilibrium state at each electrode. In this state all net reaction rates are zero, the electrochemical potentials $\tilde{\mu}$ are identical across both electrode|electrolyte interfaces, and the electrical circuit is open. Upon closing the circuit, a current flows through the system, the voltage drop between both electrodes changes, and the assumptions leading to equation (7.16) are no longer valid. This change in the cell voltage drop U^{loss} is current dependent and can be written as

$$U(j) = U^{oc} - U^{loss}(j). \tag{7.27}$$

The voltage loss can be decomposed into two contributions from the respective electrodes, called anodic and cathodic **overpotentials** $\eta_{an}(j)$ and $\eta_{cat}(j)$. In the case of galvanic cells the overpotentials at the two electrodes are both decreasing the externally usable voltage U. The anode potential is increased compared to its equilibrium value, driving electrons into the anode while the cathode potential is lowered compared to its equilibrium value driving electrons out of the cathode:

$$U(j) = E_{\text{cat}}(j) - E_{\text{an}}(j) =: E_{\text{cat}}^{\text{oc}} - E_{\text{an}}^{\text{oc}} + \underbrace{(\eta_{\text{cat}}(j) - \eta_{\text{an}}(j))}_{\leq 0} . \tag{7.28}$$

The effect of the overpotentials is shown in Figure 7.6.

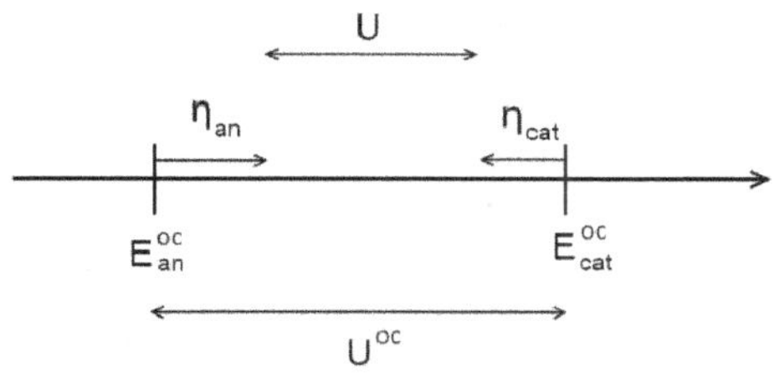

Fig. 7.6: Effect of the overpotentials in a galvanic cell.

As the reaction is no longer in equilibrium, this also means that a part of the exergy is lost during this conversion of chemical into electrical energy. The second-law efficiency for the conversion of chemical into electrical energy in the galvanic cell then becomes

$$\eta_2 = \frac{|e_{\text{el}}|}{|\Delta g_{\text{r}}^0|} = \frac{nFU}{nFU^0} = \frac{U}{U^0} = 1 - \frac{(\eta_{\text{an}}(j) - \eta_{\text{cat}}(j))}{U^0}, \tag{7.29}$$

where e_{el} is the electrical energy extracted per mole of fuels consumed. The occurrence of the overpotentials can be linked to four main mechanisms.

Other reactions: Reactions other than the desired chemical reaction that can in principle occur at both electrodes have an influence on the open-circuit voltage U^{oq} and lower its value compared to the value calculated by equation (7.16). An example is the formation of oxides at electrode surfaces. This effect is independent of the cell current.

Reaction kinetics: The reaction rate of the chemical reactions taking place at both electrodes depends on the effective energetic barrier height that has to be overcome to run the reaction. The net rate is zero in equilibrium, as both directions proceed at the same speed corresponding to identical barrier heights. As charge carrier transfers to the electrode are involved in the reaction the barrier height can be influenced by the electrode potential, and one direction can be favored. A detailed discussion of the reaction kinetics will be carried out below in Section 7.2.3.

Concentration overpotentials: If the reaction is proceeding very fast, it becomes difficult to feed the reactants and dispose of the products quickly enough to keep the concentrations of chemical species constant at the electrodes. This effect is very pronounced at high currents and acts as a cut-off for the currents achievable for a given device.

Ohmic losses: The current flowing between the electrodes leads to ohmic losses. These losses tend to align the two electrode potentials and are typically dominated by the ohmic resistance of the electrolyte and the membrane. The ohmic losses are linear with the current.

Putting together all the mechanisms described above leads to an *IU*-characteristic for a galvanic cell, as shown in Figure 7.7.

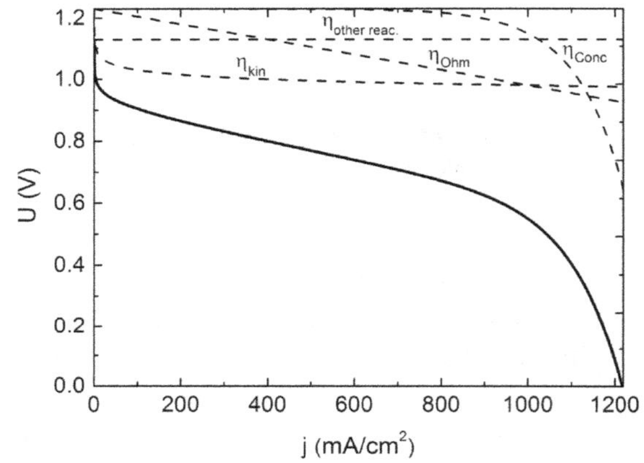

Fig. 7.7: Typical *IU*-characteristic of a galvanic cell (*black curve*) together with losses caused by the occurrence of other reactions $\eta_{\text{other reac.}}$, finite reaction kinetics η_{kin}, ohmic losses η_{Ohm}, and mass transport phenomena η_{Conc}. Realistic values for a H_2/O_2 fuel cell are depicted.

Such an *IU*-characteristic as shown in Figure 7.7 is typically used in connection with galvanic cells. A more typical characteristic for other electrical device is an *UI*-characteristic as shown in Figure 7.8.

The latter characteristic is also consistent with the sign convention used throughout this book, where a system provides work to the surroundings when the sign of the power is negative which is true for every point of the curve shown in Figure 7.8.

7.2.3 Reaction kinetics

As kinetic losses play a special role in the design of galvanic cells and determine the catalyst required, a closer look at this type of losses will now be taken. For the derivation a reaction similar to equation (7.3) is taken, i.e. a reaction where only one elec-

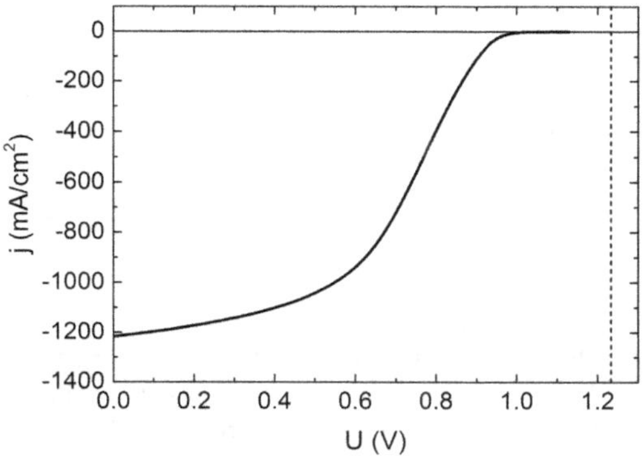

Fig. 7.8: *UI*-characteristic of the same galvanic cell shown in Figure 7.7.

tron is transferred in each reaction step. The current at an electrode is determined by the **rate constant** k of the electrochemical reaction and the activity a of the species involved:

$$j = Fkac_0. \tag{7.30}$$

The basic idea behind the concept of a rate constant is that the initial and final states of the chemical reaction are separated by an energetic activation barrier e_b. This barrier is then overcome by thermal excitation. Consequently, the rate constant of a chemical reaction is then given by

$$k = k_0 \cdot e^{-e_b/RT}, \tag{7.31}$$

with a function k_0 which is assumed to be independent of temperature. The energetic level of the initial and final state of the species can be tuned by the overpotential across the interface, as charge carriers from the electrode are involved. When the electrode potential changes, one of the two directions of the chemical half-reaction at the electrode is favored, and the other one is hindered. In a first-order expansion the total shift of the two energy levels is identical to the overpotential. Assuming that the oxidation reaction is promoted, i.e. that $\eta > 0$, in a first-order expansion in the overpotential at the electrode the barrier height for the oxidation reaction is decreasing linearly with the overpotential:

$$e_{b,ox}(\eta) =: e_{b,ox}^{eq} - \alpha F\eta. \tag{7.32}$$

The quantity $\alpha \in [0, 1]$ is the **anodic charge transfer coefficient** or **symmetry factor** which describes the symmetry of the effect of the overpotential on the energetic barrier height for the oxidation and reduction reaction, $e_{b,ox,red}$, respectively. It is typically close to $\alpha = 1/2$, corresponding to the symmetric case. Keep in mind that here oxidation and reduction reaction refers to the two possible reaction directions at one electrode and not to the reactions taking place at both electrodes. Conversely, the first-order expansion in the overpotential of the reduction reaction shows an opposite sign, as it is hindered by an increasing overpotential, and the symmetry factor is complementary:

$$e_{b,red}(\eta) =: e_{b,red}^{eq} + (1 - \alpha)F\eta. \tag{7.33}$$

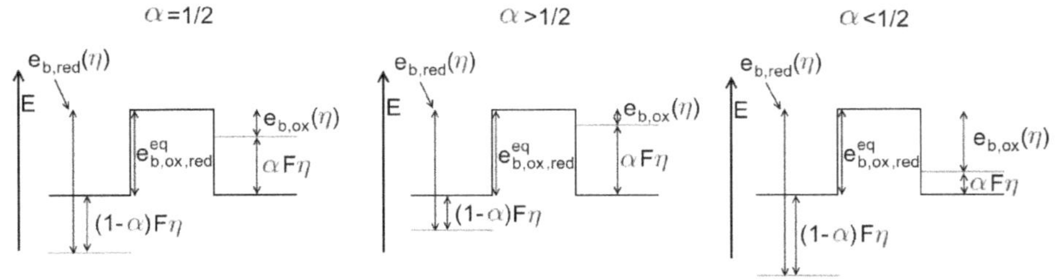

Fig. 7.9: Overpotential dependence of the energy barrier for an anodic overpotential η driving a net oxidation reaction at an electrode associated with an anodic current. The energy levels of the reduced and oxidized species are at the left- and right-hand side of the barrier, respectively.

The energy barriers and effects of an anodic overpotential for different values of α are shown in Figure 7.9.

In equilibrium the net current through the electrode surface is equal to zero. This implies that the currents calculated with equation (7.30) for both directions of the reaction have an identical absolute value, called **exchange current density** j_0, and an opposite sign:

$$j_0 = Fa_{\text{red}}c_0k_0 \cdot \exp\left(-\frac{e_{\text{b,ox}}^{\text{eq}}}{RT}\right) = -Fa_{\text{ox}}c_0k_0' \cdot \exp\left(-\frac{e_{\text{b,red}}^{\text{eq}}}{RT}\right), \tag{7.34}$$

where $a_{\text{red,ox}}$ are the activities of the reduced and oxidized species and $e_{\text{b,ox,red}}^{\text{eq}}$ the equilibrium barrier heights for the oxidation and reduction reaction, respectively. The exchange current density is a constant, depending on the electrode material, especially the catalyst. For a good performance, i.e. low kinetic losses, an electrode material with a high j_0 has to be chosen. The total $j\eta$-characteristic of a single electrode can then be obtained using equations (7.31)–(7.34):

$$\begin{aligned} j(\eta) &= zFk_{\text{ox}}(\eta)a_{\text{red}}c_0 - Fk_{\text{red}}(\eta)a_{\text{ox}}c_0 \\ &= Fa_{\text{red}}c_0 \cdot k_0 \cdot \exp\left(-\frac{e_{\text{b,ox}}(\eta)}{RT}\right) - Fa_{\text{ox}}c_0 \cdot k_0' \cdot \exp\left(-\frac{e_{\text{b,red}}(\eta)}{RT}\right) \\ &= Fa_{\text{red}}c_0 \cdot k_0 \cdot \exp\left(-\frac{e_{\text{b,ox}}^{\text{eq}} - \alpha F\eta}{RT}\right) - Fa_{\text{ox}}c_0 \cdot k_0' \cdot \exp\left(-\frac{e_{\text{b,red}}^{\text{eq}} + (1-\alpha)F\eta}{RT}\right) \\ &= j_0\left(\exp\left(\alpha\frac{F}{RT}\eta\right) - \exp\left(-(1-\alpha)\frac{F}{RT}\eta\right)\right). \end{aligned} \tag{7.35}$$

This equation is called the **Butler–Volmer equation.** It describes the effect of an overpotential on the current at a single electrode. Anodic and cathodic partial currents as well as the total current as a function the overpotential are depicted in Figure 7.10.

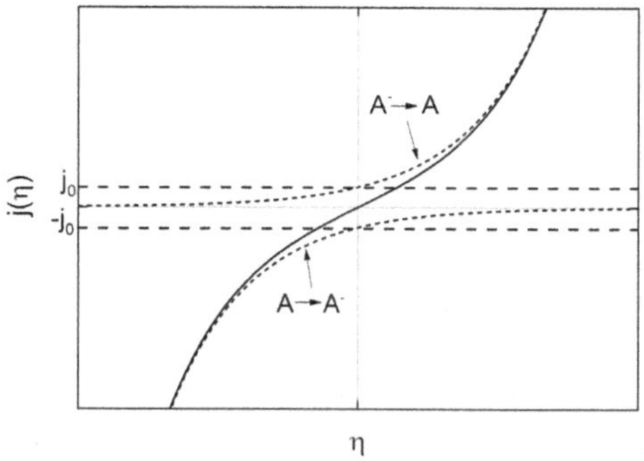

Fig. 7.10: Current overpotential characteristics at a single electrode according to the Butler–Volmer equation (7.35) for $\alpha = 0.5$ (*solid curve*) and the anodic and cathodic partial currents corresponding to both summands in equation (7.35) (*dashed curves*).

Two Butler–Volmer equations (equation (7.35)) with different values of j_0 and α characterize the kinetics at the two electrodes of an electrochemical cell. From these two current-overpotential characteristics the total effect of the kinetics on the cell voltage can then be obtained. The oxidation current at the anode and the reduction current at the cathode have an identical absolute value and an opposite sign, as an identical current flows through all components connected in series in an electrical circuit. At both electrodes, curves as in Figure 7.10 can be drawn centered at the respective equilibrium potentials. The overall effect of the reaction kinetics at both electrodes at a given current is shown in Figure 7.11.

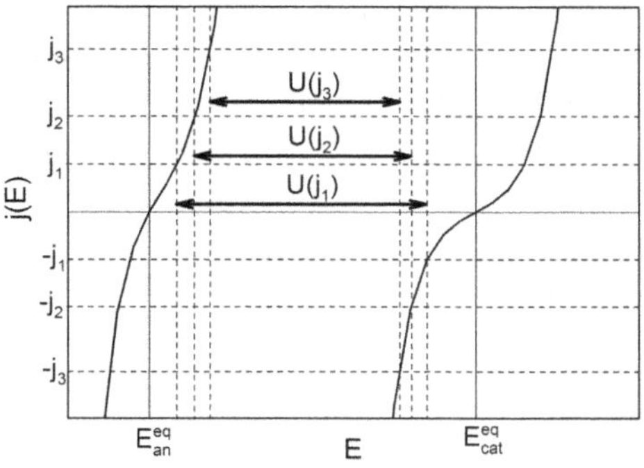

Fig. 7.11: Current overpotential relation according to the Butler Volmer equation (equation (7.35)) for anode (*left*) and cathode (*right*). The cell voltages $U(j_{1,2,3})$ for three distinct current levels $j_{1,2,3}$ are depicted.

7.2.4 Fuel cells

A fuel cell is an open galvanic cell where an identical electrolyte at both sides of the membrane is constantly fed with two different chemical species, and the reaction products are continuously disposed of. By far the most common type of cell is the

H_2/O_2 fuel cell, where the two half-reactions are given by

$$\begin{aligned}&\text{Cathode:} \quad O_2 + 4H^+ + 4e^- \rightarrow 2H_2O \ ; \ E^0 = 1.23\ \text{V vs. SHE}\\ &\text{Anode:} \quad 2H_2 \rightarrow 4H^+ + 4e^- \ ; \ E^0 = 0.00\ \text{V vs. SHE.}\end{aligned} \tag{7.36}$$

In a low temperature H_2/O_2 fuel cell the electrolyte is a **proton exchange membrane** (PEM), which acts at the same time as a diffusion barrier for the reactants. It exclusively allows the crossing of protons. The design of a fuel cell is closely linked to the causes of the overpotential discussed in Section 7.2.2. In the following these design criteria are discussed for H_2/O_2 fuel cells. To minimize kinetic losses by maximizing j_0 a suitable catalyst has to be used as electrode material. In H_2/O_2 fuel cells this catalyst is typically platinum at both sides. As platinum is an expensive and limited material, the amount of catalyst used has to be kept at a minimum. At the same time the internal surface area of the electrode has to be as large as possible to achieve a high output power. Both necessities lead to the usage of platinum nanoparticles (diameter ≈ 5 nm) on porous conductive carbon fiber backbones as electrodes, which are interpenetrated by the polymer electrolyte. The nanoparticles themselves as well as the backbone then have a high surface to volume ratio. To minimize ohmic losses the fuel cell is built in a very compact manner, and the active layer consisting of both electrodes and the solid H^+ conducting electrolyte, which also acts as the membrane, has an overall width of no more than several 10 μm. Finally, it is important to guarantee a stable and uniform transport of reactants and products. This is achieved by meandering gas channels parallel to the extension of the electrode cut into the metal plates that give the cell mechanical stability. The gas then passes a diffusion medium, i.e. a porous conductive medium that assures a uniform gas diffusion to the electrodes. The reaction itself requires a three-phase boundary between the gaseous reactants, the platinum nanoparticles, and the polymer electrolyte. In addition to the gas transport the metal plates act as electrical leads and are called **bipolar plates**, as in a stack of fuel cells, where many cells are connected in series, they act as electron donors on the one side and as electron acceptors on the other side. A sketch of an H_2/O_2 fuel cell is shown in Figure 7.12.

7.2.5 Nonrechargeable batteries

By far the most commonly used direct converters of chemical into electrical energy are batteries. In this section nonrechargeable batteries will be discussed, while rechargeable batteries will be discussed in Section 11.3.1 in the context of exergy storage. The most widely used type of nonrechargeable battery is the **alkaline battery**. In this type of battery a zinc (Zn) electrode as anode and a manganese dioxide (MnO_2) electrode as cathode are each immersed in an aqueous strongly alkaline potassium hydroxide (KOH) electrolyte. While the details are a little more complicated, the following mini-

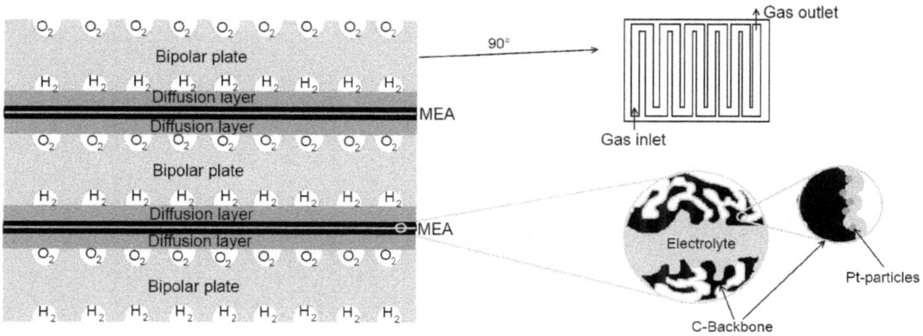

Fig. 7.12: Fuel cell stack consisting of several membrane electrode assemblies (MEAs) connected in series via bipolar metal plates (*left*). The meandering channels in the bipolar plates and the a detailed view of the MEAs is shown on the *right*.

mal two half-reactions and resulting full reaction can be used:

$$\text{Cathode}: \quad Zn + 2OH^- \rightarrow ZnO + H_2O + 2e^-; \; E^0 = -1.28 \text{ V vs. SHE}$$
$$\text{Anode}: \quad 2MnO_2 + 2H_2O + 2e^- \rightarrow 2MnO(OH) + 2OH^-; \; E^0 = 0.15 \text{ V vs. SHE}$$
$$\Rightarrow \quad \text{Full reaction}: \quad Zn + 2MnO_2 + H_2O \rightarrow 2MnO(OH) + ZnO \qquad (7.37)$$

With this, the full standard cell voltage is $E^0 = 1.43$ V. In a real cell this value is typically a little higher and lies above $E = 1.5$ V. The chemical species that can pass from one electrode to the other, thus internally closing the electric circuit, is OH^-. Both half-reactions involve changes of the electrodes themselves. The current voltage characteristic at a given time is similar to the one shown in Figure 7.8. The battery is a closed system and is characterized by its **capacity** C, which is the total amount of charge the battery can deliver before the voltage drops below a predefined value. This capacity is not to be confused with an electronic capacitance and is typically given in ampere hours (Ah) rather than coulomb (C). As all reactants except of water are in the solid-state phase, their activity does not change significantly upon discharge. The nonreacted electrode surfaces in contact with the electrolyte, however, shrink in size, leading to higher kinetic losses for a given current. Together with the loss of water, this effect leads to a gradual drop in the usable cell voltage with the total amount of charges extracted. For a given amount of extracted charges, this voltage drop is more pronounced when the battery is discharged faster. In addition, the total exergy extracted is further lowered for a fast discharge, as this happens with a lower second-law efficiency, as evident from Figure 7.8, while the total amount of exergy stored in the battery is fixed.

8 Solar energy

Solar radiation constitutes the most abundant exergy source, and solar energy conversion is therefore essential for the transformation of our energy supply into a sustainable one. As the first step of a detailed study of the different solar energy conversion processes, the properties of sunlight will be discussed in this chapter.

8.1 Properties of solar irradiation

The radiation coming from the sun can be well approximated by heat transferred by an ideal black body at a temperature $T_{sun} \approx 5770$ K. To determine the spectral distribution of this heat radiation, the photon field emitted by the sun has to be treated as a boson gas. The spectral energy volume density $d(U/V)/dE$ of a boson ensemble at temperature T can be calculated using the spectral density of states $D(E)$ and the Bose–Einstein distribution $f(E,T)$, as discussed in Section B.1 of the appendix (see equation (B.10)):

$$\begin{aligned} \frac{d(U/V)}{dE} &= D(E) \cdot f(E,T) \cdot E = \frac{8\pi}{(hc)^3} E^2 \cdot \frac{1}{e^{E/kT} - 1} \cdot E \\ &= \frac{8\pi}{(hc)^3} \frac{E^3}{e^{E/kT} - 1}, \end{aligned} \tag{8.1}$$

where c is the velocity of light. This law was originally derived by Max Planck and is one of the foundations for the development of quantum mechanics. It is also the first time the term h for the quantum of action was used, it is short for the German word *Hilfsparameter* which roughly translates as “auxiliary parameter”. It has a value of $h = 4.14 \cdot 10^{-15}$ eVs. With equation (8.1), the spectral energy density that passes a given area on earth perpendicular to the light path per unit time, i.e. the spectral light intensity on the earth, can be calculated using the geometrical construction for the infinitesimal volume, shown in Figure 8.1a, and the geometry of the sun/earth system, shown in Figure 8.1b.

Using Figure 8.1a it can be seen that the expression $dV = c \cdot dt \cdot dA$ holds. Thus, equation (8.1) can be reformulated as a function of the spectral light intensity, i.e. the power of the radiation per unit area, $dI := d(P/A)$ in the following way:

$$\frac{dI}{dE} := \frac{d(P/A)}{dE} = c \cdot \frac{d(U/V)}{dE} = \frac{8\pi}{h^3 c^2} \frac{E^3}{e^{E/kT} - 1}. \tag{8.2}$$

Equation (8.2) only holds if the photon field is homogeneous, i.e. if the light rays are parallel. For a different geometry this value is changed, and for the following discussion of this change the homogeneous intensity spectrum as given in equation (8.2)

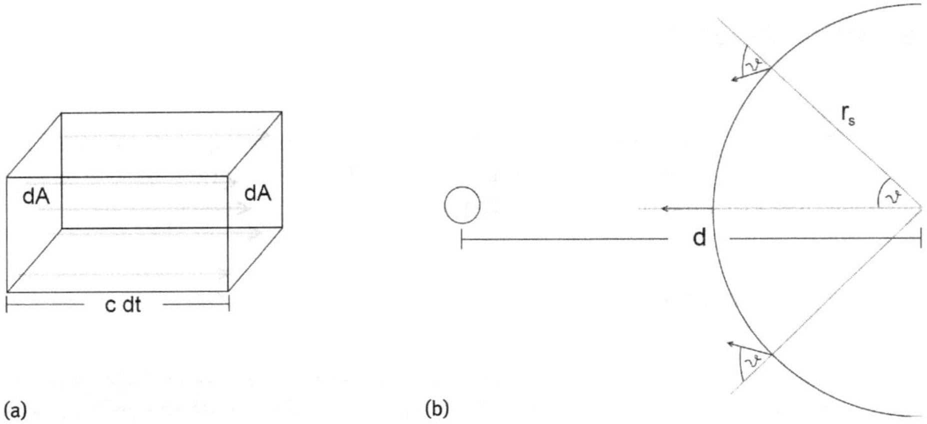

Fig. 8.1: Geometrical considerations for the calculation of the sunlight intensity at the earth: (a) conversion of energy density into intensity; (b) Geometry of the sun/earth system.

will be denoted by $(dI/dE)_{\text{hom}}$.

To account for the geometry of the sun/earth system the radiation reaching the earth emitted from an area dA_s at the solar surface is considered. With respect to the area normal at the surface element, the direction of the earth is characterized by the angle ϑ, as shown in Figure 8.1b, and the size of the area element as seen from earth $A_{\text{s,projected}}$ is given by its projection on the plane perpendicular to the sun/earth connection:

$$dA_{\text{s,projected}} = dA_\text{s} \cdot \cos(\vartheta). \tag{8.3}$$

As the radius of the sun is much smaller than the distance between the sun and the earth, the distance between the individual areas at the solar surface and the earth does not change significantly and can be approximated by the average distance d. Consequently, each of the area elements at the solar surface emits a spherical wave whose spectral intensity in a distance d and the direction of the earth, i.e. at an angle ϑ to the area normal, is given by

$$\begin{aligned} d\left(\frac{dI}{dE}\right) &= \left(\frac{dI}{dE}\right)_{\text{hom}} \cdot \frac{dA_{\text{s,projected}}}{4\pi d^2} = \left(\frac{dI}{dE}\right)_{\text{hom}} \cdot \frac{dA_\text{s}}{4\pi d^2}\cos(\vartheta) \\ \Rightarrow \quad \frac{d^2 I}{dE \cdot dA_\text{s}} &= \left(\frac{dI}{dE}\right)_{\text{hom}} \cdot \frac{\cos(\vartheta)}{4\pi d^2}. \end{aligned} \tag{8.4}$$

The total spectral intensity of the solar radiation reaching the earth can then be calculated by integrating equation (8.4) over the entire solar surface visible from the earth and setting the distance to $d = 1\ \text{AU} \approx 1.5 \cdot 10^{11}$ m, the average distance between the sun and the earth:

$$
\begin{aligned}
\left(\frac{dI}{dE}\right)_{\text{earth}} &= \int\limits_{\text{vis.sun}} \frac{d^2I}{dE\cdot dA_s}dA_s \\
&= \frac{1}{4\pi d^2}\cdot\left(\frac{dI}{dE}\right)_{\text{hom}} \int\limits_0^{2\pi}\int\limits_0^{\pi/2} r_s^2 \sin(\vartheta)\cos(\vartheta)d\vartheta\, d\varphi \\
&= \frac{r_s^2}{4\pi d^2}\cdot\left(\frac{dI}{dE}\right)_{\text{hom}} \int\limits_0^{2\pi}\int\limits_0^{1} \cos(\vartheta)d(\cos(\vartheta))d\varphi \\
&= \frac{r_s^2}{4\pi d^2}\cdot\left(\frac{dI}{dE}\right)_{\text{hom}} \int\limits_0^{2\pi} 1/2 d\varphi \\
&= \frac{r_s^2}{4d^2}\cdot\left(\frac{dI}{dE}\right)_{\text{hom}} =: \frac{1}{4f}\cdot\left(\frac{dI}{dE}\right)_{\text{hom}} = \frac{1}{f}\frac{2\pi}{h^3c^2}\frac{E^3}{e^{E/kT}-1}.
\end{aligned}
\tag{8.5}
$$

This expression can also be found if the projection of the sun visible from the earth is taken to be a homogeneously emitting circle. The integration over the surface area of the half sphere is then simply replaced by the integration over the area of the circle, which is πr_s^2. For the **geometrical factor** f of the sun/earth system, the following definition is used:

$$f := \frac{d^2}{r_s^2} \quad \Rightarrow \quad f \approx 46\,200. \tag{8.6}$$

A comparison of the real solar spectrum just outside the atmosphere of the earth and the spectrum calculated using equation (8.5) is shown in Figure 8.2.

The reasonable agreement between the two curves shown in Figure 8.2 affirms that the approximative treatment of the sun as a black body is well justified.

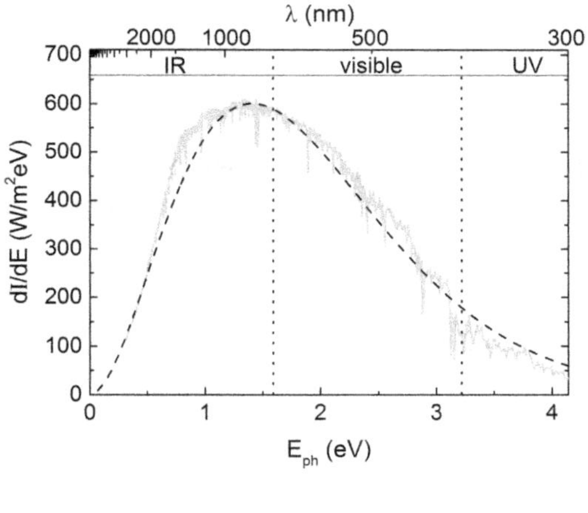

Fig. 8.2: The solar spectrum measured just outside of the atmosphere of the earth (*gray curve*) and the spectrum calculated via equation (8.5) (*black dashed curve*).

8.1.1 Stefan–Boltzmann law

To calculate the total light intensity reaching the earth from the sun, the spectral light intensity given in equation (8.5) has to be integrated again, this time over all photon energies:

$$\begin{aligned} I_{\text{earth}} &= \frac{1}{f}\frac{2\pi}{h^3c^2}\int_0^\infty \frac{E^3}{e^{E/kT}-1}dE \underbrace{=}_{y:=E/kT} = \frac{1}{f}\frac{2\pi(kT)^4}{h^3c^2}\underbrace{\int_0^\infty \frac{y^3}{e^y-1}dy}_{=\pi^4/15} \\ &= \frac{1}{f}\frac{2\pi^5(kT)^4}{15h^3c^2} =: \frac{1}{f}\sigma T^4. \end{aligned} \tag{8.7}$$

Apart from the factor $1/f$, which is a constant in the sun/earth system, this expression is the **Stefan–Boltzmann law**, which most notably states that the total intensity of radiation emitted from a black body at temperature T is proportional to T^4. The proportionality constant $\sigma = 5.67 \cdot 10^{-8}\ \text{W}/(\text{m}^2\text{K}^4)$ is called the **Stefan–Boltzmann constant.** With equation (8.7), the total intensity of the sunlight just outside the atmosphere of the earth can be obtained:

$$I_{\text{earth}} = \frac{1}{f}\sigma T_{\text{sun}}^4 = 1.35\ \text{kW/m}^2. \tag{8.8}$$

This value is called the **solar constant.** Due to variations in the distance between the sun and the earth, this value changes over the year between a maximum of 1.4 kW/m^2 in early January and a minimum of 1.3 kW/m^2 in early July. It is thus locked to the seasons on the earth, leading to a small weakening of the seasonal effects on the northern hemisphere and a small increase on the southern hemisphere. If the atmosphere is neglected, this leads to an annual average incident energy of 5.9 MWh/(m^{2}a), i.e. 760 EWh/a on the entire earth. This value is very large compared to the 13.4 Gtoe = 0.16 EWh/a primary energy consumption of humanity in the year 2012 mentioned in Section 1.1.2.

8.1.2 Exergy of the solar irradiation

The influx of exergy from the sun can be calculated with the following expression for the volume specific entropy s of black body radiation (for a more detailed derivation see Section B.1.3 of the appendix):

$$\begin{aligned} s &= \frac{S}{V} = \frac{16\sigma}{3c}T^3 \\ \Rightarrow \frac{\dot{S}}{A} &= c \cdot \frac{S}{V} = \frac{16\sigma}{3}T^3. \end{aligned} \tag{8.9}$$

With the latter equation, the stream of entropy from the sun to the earth can be calculated with the identical geometrical considerations as in equations (8.2)–(8.5). The difference between equations (8.2) and (8.5) reveals that the geometry of the sun/earth system leads to a factor of $1/(4f)$ compared to the case of a homogeneous field. This means that the total entropy stream density to the earth is given by

$$\frac{\dot{S}}{A} = \frac{4\sigma}{3f} T_{\text{sun}}^3. \tag{8.10}$$

As the anergy content of any energy crossing the border of a system, in this case the earth, is given by $T_{\text{am}}S$, as derived in Section 3.2, the total exergy stream density from the sun is given by

$$\frac{\dot{\Psi}}{A} = I_{\text{earth}} - T_{\text{am}}\frac{\dot{S}}{A} = \frac{1}{f}\sigma T_{\text{sun}}^4 - \frac{4}{3f}\sigma T_{\text{sun}}^3 \cdot T_{\text{am}} = \left(1 - \frac{4T_{\text{am}}}{3T_{\text{sun}}}\right) \cdot I_{\text{earth}} = 1.27\,\text{kW/m}^2. \tag{8.11}$$

The solar irradiation is hence mostly exergy, and the maximum first-law conversion efficiency $\eta_{1,\text{max}}$ of any device converting solar energy to electrical or mechanical energy is given by

$$\eta_{1,\text{max}} = \left(1 - \frac{4T_{\text{am}}}{3T_{\text{sun}}}\right) = 0.93. \tag{8.12}$$

There is a spectral distribution of the entropy stream per unit area $\dot{S}/A$ of black body radiation. It is derived in Section B.1.3 of the appendix and given by

$$\frac{d(\dot{S}/A)}{dE} = \frac{1}{T_{\text{sun}}}\left(\frac{dI}{dE}\right)_{\text{earth}} - k\frac{1}{f}\frac{2\pi}{h^3c^2}E^2 \ln\left(1 - e^{-E/kT_{\text{sun}}}\right). \tag{8.13}$$

With equation (8.13) the spectral exergy content of the photon field can be calculated:

$$\begin{aligned}\frac{d(\dot{\Psi}/A)}{dE} &= \left(\frac{dI}{dE}\right)_{\text{earth}} - T_{\text{am}} \cdot \frac{d(\dot{S}/A)}{dE} \\ &= \left(1 - \frac{T_{\text{am}}}{T_{\text{sun}}}\right)\left(\frac{dI}{dE}\right)_{\text{earth}} - kT_{\text{am}} \cdot \frac{1}{f}\frac{2\pi}{h^3c^2}E^2 \ln\left(1 - e^{-E/kT_{\text{sun}}}\right).\end{aligned} \tag{8.14}$$

The spectral distribution of the exergy stream coming from the sun is relevant for any solar energy-converting device which only makes use of a part of the solar spectrum. Photovoltaic cells are the most important example for such devices, and will be discussed in detail in Chapter 10. The spectrum of the exergy content of the solar radiation is shown in Figure 8.3.

The bulk of the anergy is transported via photons with a relatively low energy. For photons with an energy above ≈ 1 eV, the spectral exergy content is approximately identical to the average exergy content of the entire spectrum, as can be seen by comparing the exergy content curve with the *dashed line* in Figure 8.3.

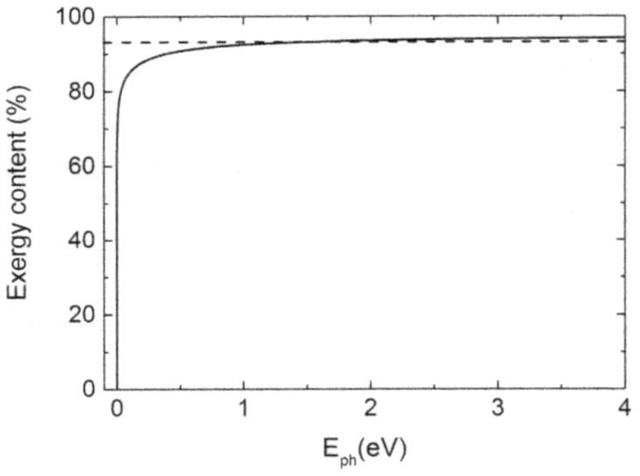

Fig. 8.3: Spectral distribution of the exergy content of the black body solar radiation. The *dashed line* represents the spectral average.

8.2 Influence of the atmosphere

Intensity and spectral composition of the incoming sunlight is influenced by the atmosphere of the earth via scattering and absorption processes. Due to the scattering processes in the atmosphere, the light reaching the surface of the earth is not all coming directly from the direction of the sun. If that were the case, the sky would look the way it does on the moon, and shadows of objects would be pitch black. The light scattered at one or multiple scattering centers is incident from the direction of the last scattering center involved in the scattering process. As the scattering centers are homogeneously distributed throughout the atmosphere the scattered light is diffuse. It is called **indirect radiation**, as opposed to the unscattered **direct radiation**. The total intensity I_{tot} of the sunlight at a given spot can then be decomposed in the following way:

$$I_{\mathrm{tot}} = I_{\mathrm{direct}} + I_{\mathrm{indirect}}. \tag{8.15}$$

The fraction of indirect light varies strongly with the latitude. In central Europe for example only 40 % of the total solar irradiation is direct on average, whereas this value increases to 70–80 % close to the equator. This fact has strong implications for the feasibility of solar energy conversion technologies. In general, especially concentrating devices require mostly direct radiation which makes them unfavorable in higher latitudes.

The scattering processes can be classified into **Rayleigh scattering**, where the scattering centers have a size $r \ll \lambda$, **Mie scattering**, with $r \approx \lambda$, and **geometric scattering** with $r \gg \lambda$. The cross section of the geometrical scattering for example at clouds does not show a spectral dependence, and the cross section of the Mie scattering does so only slightly. For the Rayleigh scattering the cross section is proportional to E^4 and leads to a cut-off of the solar radiation reaching the surface of the earth at wavelengths below 200 nm (6.2 eV). It is also responsible for the blue appearance of the sky.

Molecules in the atmosphere, like most famously ozone (O_3), water (H_2O) and carbon dioxide (CO_2), have discrete spectral bands, where they are excited and are consequently strongly absorbing the incoming light. The molecular absorption leads to the absorption of discrete energy bands in the photon spectrum. Especially water shows prominent absorption features in the visible and near-infrared part of the spectrum which is most prominent in the solar irradiation. Ozone, on the other hand, provides a cut-off for wavelengths below ≈ 350 nm (3.5 eV), i.e. in a wavelength region which is very harmful to many species, including humans. All the mentioned intensity losses scale with the thickness of the atmosphere passed by the solar irradiation. Their effect is captured in standardized solar spectra by assigning a standardized atmospheric column, the so-called **air mass (AM)**, to the spectra. AM1 corresponds to the sun standing at the zenith. Geometric scattering is neglected in the standardized spectra, which means that a cloud-free sky and unpolluted air is assumed. As the sun is in general not standing at the zenith, the effect of the atmosphere would be underestimated using AM1 spectra. The standard spectrum used is therefore AM1.5, corresponding to a 1.5 times longer path through the atmosphere than for the sun at the zenith. It corresponds to an incident angle of 48.2°. AM0 and A1.5 are compared in Figure 8.4.

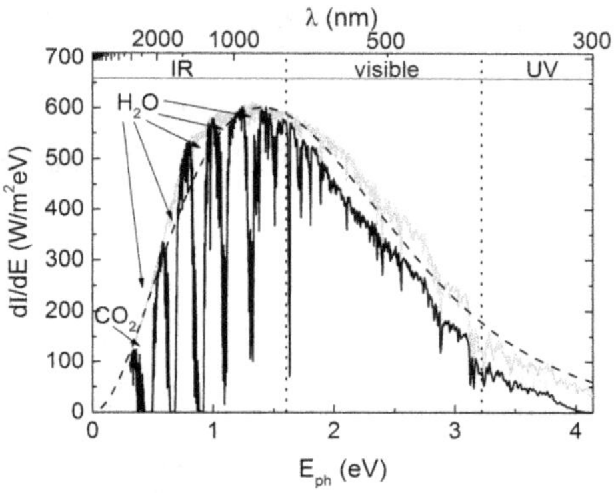

Fig. 8.4: AM1.5 spectrum (*solid black curve*) together with the AM0 spectrum, i.e. the radiation of the sun reaching the earth just outside the atmosphere of the earth (*gray curve*), and the spectrum calculated with equation (8.5) (*black dashed curve*). The infrared, visible, and UV region are marked (*dotted lines*) as well as the molecules corresponding to the absorption bands.

The total light intensity reaching the surface of the earth after 1.5 air masses is reduced compared to the solar constant and has a value of $I_{AM1.5} \approx 1\ \text{kW/m}^2$, corresponding to 4.4 MWh/(m²a) or 560 EWh/a. A part of the solar radiation reaching the lower atmosphere is reflected by the atmosphere and the surface of the earth, and thus not participating in any energy conversion process. This effect accounts for an energy loss of ≈ 30 % over the entire spectral range and is typically referred to as **albedo**. The average annual incident solar energy at the surface of the earth is then ≈ 3 MWh/(m²a) or ≈ 400 EWh/a for the entire earth. This means that while the atmosphere significantly diminishes the total incident solar radiation, this effect is not affecting the abundance of solar energy.

8.2.1 The greenhouse effect

Arguably, an even more important effect of the molecules in the atmosphere of the earth than the attenuation of the incoming solar radiation is their action on the thermal radiation emitted by the earth into space. In the spectral range of this thermal radiation ($\lambda \approx$ 10–20 µm, $E_{ph} \approx$ 50–100 meV) the molecular absorption is far more pronounced than in the visible and near-infrared relevant for solar irradiation. The molecules responsible are mainly H_2O and CO_2. This selective absorption (see Section 9.1.1 below), i.e. stronger absorption in the spectral range of the radiation emitted from the earth than absorbed from the sun, leads to a heating of the earth and is known as the **greenhouse effect.** Without the influence of the atmosphere the temperature of the earth T_{earth} modeled as a black body can be calculated using the Stefan–Boltzmann law (equation (8.7)):

$$\frac{1}{f}\sigma T_{sun}^4 = 4\sigma T_{earth}^4$$
$$\Rightarrow \quad T_{earth} = (4f)^{-1/4} T_{sun} = 279\ \text{K} \approx 6\ ^\circ\text{C}, \tag{8.16}$$

where the factor 4 on the right-hand side reflects the fact that the projection of the earth seen from the sun is πr_{earth}^2, while the thermal emission is from the entire surface of the spherical earth $4\pi r_{earth}^2$. This temperature decreases further to T_{earth} = 254 K = −19 °C if the albedo effect, i.e. the reflection of sunlight, is taken into account. Due to the greenhouse effect the global average temperature is increased by 33 °C to T_{earth} = 287 K = 14 °C, as illustrated in Figure 8.5.

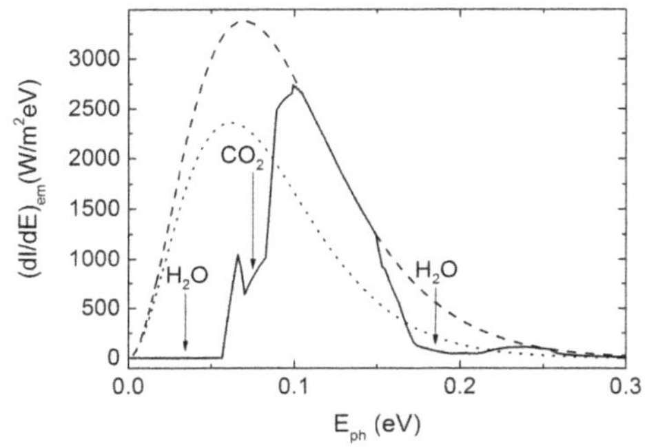

Fig. 8.5: Black body radiation of the earth without atmosphere at T_{earth} = 254 K (*dotted curve*), together with a realistic emission spectrum (*solid curve*) and the corresponding black body spectrum at T_{earth} = 287 K (*dashed curve*). The area under the *solid* and *dotted curves* are both the same and are identical to the time average of the incoming solar AM1.5 radiation with albedo.

In Figure 8.5 the *dotted line* is the black body radiation the earth would emit as an ideal black body at a temperature of −19 °C. The area under this curve is then identical to the absorbed solar irradiation. It is then also identical to the actually emitted thermal radiation from the earth with atmosphere, i.e. to the area under the *solid curve*. The *dashed curve*, in turn, is then the black body spectrum at the temperature increased by the greenhouse effect, i.e. at 14 °C.

The overall greenhouse effect is dominated by the natural greenhouse effect, i.e. by the atmospheric composition at preindustrial levels. As CO_2 is an important greenhouse gas, the man-made increase of its concentration in the atmosphere from 280 ppm (parts per million) in preindustrial times to ≈ 400 ppm today in the course of roughly a century has an increasing effect on the global average temperature. This additional contribution to the overall greenhouse effect is the **anthropogenic greenhouse effect**. It is this additional effect which is playing a central role in the public discourse concerning global warming. Note that the official international goal set by the United Nations via the IPCC (Intergovernmental Panel on Climate Change) to keep the man-made global warming below 2 °C means that the anthropogenic component should not exceed 6 % of the overall greenhouse effect, which then amounts to an overall warming of 35 °C. While the exact contribution of the CO_2 to the overall greenhouse effect is not easy to determine, it is certainly one of the major contributors. The comparison of the aim to not increase the overall greenhouse effect by more than 6 % to the already mentioned man-made increase of 40 % of the CO_2 concentration compared to preindustrial levels thus offers a valuable perspective when considering the reality of man-made global warming.

8.3 Photosynthesis

By far the most common and most important solar energy converters on our planet are plants, or more precisely, green plants. These life forms use the aforementioned abundance of solar energy to convert water and atmospheric CO_2 to organic compounds and oxygen in a process called **photosynthesis**. These organic compounds are then both the material and the energetic basis for nearly all other life forms. Furthermore, the oxygen produced is the only source of molecular oxygen in the atmosphere, again a necessary prerequisite for the existence of all life forms which are not themselves able to perform photosynthesis.

In green plants, the photosynthetic reaction takes place in organelles called **chloroplasts**, which are responsible for the green appearance of the plants. The overall chemical reaction during photosynthesis depends on the final organic compound and for the example of glucose takes the following form:

$$6CO_2 + 6H_2O \rightarrow C_6H_{12}O_6 + 6O_2: \quad \begin{aligned} \Delta h_r^0 &= 2810\,\mathrm{kJ/mol} \\ \Delta g_r^0 &= 2730\,\mathrm{kJ/mol}, \end{aligned} \tag{8.17}$$

where the necessary energy input is provided by the solar photons. This reaction does not take place in a single step, but runs through a complex set of intermediate states and species. It can, however, roughly be broken down to two main parts. In the first part, the **light reaction**, the energy input of the solar photons is used to split water and form chemical energy carriers such as NADPH (nicotinamide adenine dinucleotide phosphate in its reduced form). In the second part, the **dark reaction**, these energy carriers are utilized to convert CO_2 to sugar:

$$\begin{aligned}&12H_2O + 12NADP^+ \rightarrow 6O_2 + 12NADPH + 12H^+: && \text{light reaction,}\\ &6CO_2 + 12NADPH + 12H^+ \rightarrow C_6H_{12}O_6 + 6H_2O + 12NADP^+: && \text{dark reaction,}\end{aligned} \tag{8.18}$$

where $NADP^+$ is the oxidized low-energy form of NADPH. In addition to the $NADP^+$/NADPH pair, the ADP/ATP-pair (adenosin triphosphate/adenosin diphosphate) also plays a role in the overall reaction as chemical energy carrier. It does, however, not contribute to the overall atom balance of the reactions, and is thus neglected here. The enzymes catalyzing the light reaction are located in a membrane separating two compartments within the chloroplasts. A simplified picture of the membrane and the relevant reaction sites is shown in Figure 8.6.

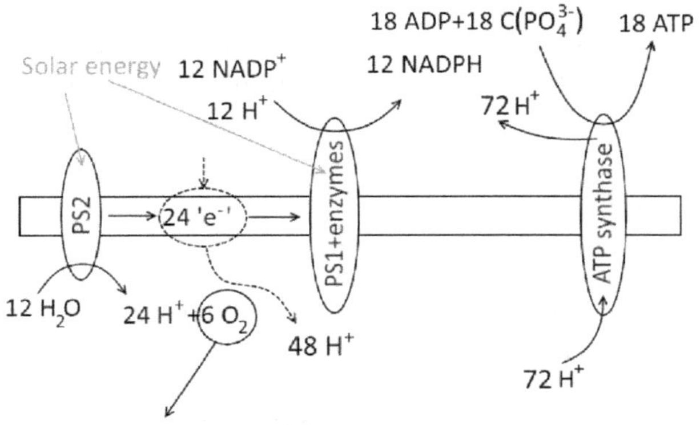

Fig. 8.6: Simplified schematic of the membrane within a chloroplast. The three relevant chemical reaction centers, photosystem II (PS2), photosystem I, and a collection of enzymes (PS1+enzymes), and the enzyme ATP synthase are shown, where the two photosystems are light activated. The electron transfer from PS2 to PS1 is in itself a complex process involving additional molecules and reaction centers, as there are no free electrons within the membrane.

The light reaction is driven by two photosystems: **photosystem II** (PS2) and **photosystem I** (PS1), which are connected in series. For the synthesis of one molecule of glucose, photosystem II absorbs 24 photons, via chlorophyll molecules which are photochemically excited. The electrons in the excited state are transferred to photosystem I by molecular charge shuttles in a series of steps. The left-behind, oxidized PS2 ($PS2^+$) is then reduced by electrons from water molecules, which are, again with the aid of further catalysts and in a complex series of steps, oxidized to molecular oxygen:

$$12H_2O + 24PS2^+ \rightarrow 6O_2 + 24H^+ + 24PS2. \tag{8.19}$$

The $24e^-$ are transported to the oxidized state of PS1. During this transfer process the electrons lose part of their energy to drive a proton pump, which in turn increases the proton gradient across the membrane. By a light induced process again involving chlorophyll the electrons are then excited a second time in PS1, before they participate in the $NADP^+$ reduction reaction. This second excitation requires an input of another 24 solar photons:

$$12NADP^+ + 12H^+ + 24e^- \rightarrow 12NADPH. \tag{8.20}$$

Thus each electron was photoexcited twice, leading to an overall consumption of 48 photons per 12 NADPH molecules. The proton concentration gradient is used to drive the enzyme ATP synthase, which produces ATP from ADP and some phosphate containing complex $C(PO_4^{3-})$. The NADPH and ATP above the membrane in Figure 8.6 then participate in the dark reaction.

8.3.1 Conversion efficiency

The external energy input to the photosynthesis reaction are a total of 48 solar photons per synthesized $C_6H_{12}O_6$ molecule. It follows from equation (8.17) that each glucose molecule has a chemical energy content of $\Delta h_r^0/N_A = 29.1$ eV and a chemical exergy content of $\Delta g_r^0/N_A = 28.3$ eV. This means that the energy output per absorbed photon is 0.61 eV, and the exergy output per absorbed photon is 0.59 eV. The absorption process itself is mainly achieved by the complex dye molecules **chlorophyll-a/b**, which are assembled in a so-called **light harvesting complex** (LHC). These molecules absorb mostly in the red and blue part of the spectrum. However, the higher energy excitation in the blue spectral region quickly degrades to a usable energy of 1.8 eV and a corresponding exergy of $\Psi_{ph} = 1.7$ eV. This means that the first- and second-law conversion efficiencies from the absorbed photons to the end product are rather high with $\eta_1 = 0.34$ and $\eta_2 = 0.35$, respectively. However, for the overall efficiency, further loss mechanisms have to be taken into account. This includes the reflection of parts of the solar spectrum, mainly in the green spectral region, saturation effects (the photosynthesis works most efficiently at ca. 5 % of the maximal intensity of the solar irradiation), and energy necessary to maintain the organism of the plant, as for example the metabolism of the plant. Altogether the theoretical maximum efficiency is then only ca. 4.5 %. Under realistic conditions only ca. 1 % of the energy carried by the solar irradiation is transformed into biomass. When this biomass is further converted to chemical fuels, the overall efficiency for the conversion from solar energy to these bio fuels is then very low indeed.

9 Solarthermal energy conversion

In a solarthermal energy collector the incident solar radiation is used to heat up an absorber, which then transfers the collected energy in form of heat to a (typically liquid) medium (thermal fluid) constantly streaming along its back side, as shown in Figure 9.1.

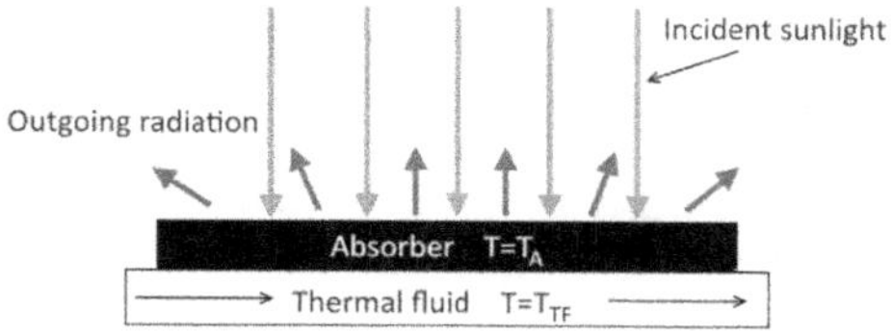

Fig. 9.1: Schematic of a solarthermal energy converter.

Under steady-state conditions, the total amounts of energy per unit time and area entering and leaving the collector are identical, i.e. $\dot{q}_{\text{in}} = \dot{q}_{\text{out}}$. In an efficient solarthermal converter the energy leaves the absorber mostly by means of transfer to the thermal fluid. Heat losses occur on the illuminated (front) and the nonilluminated (back) side. To minimize the heat losses the absorber is typically built as schematically shown in Figure 9.2.

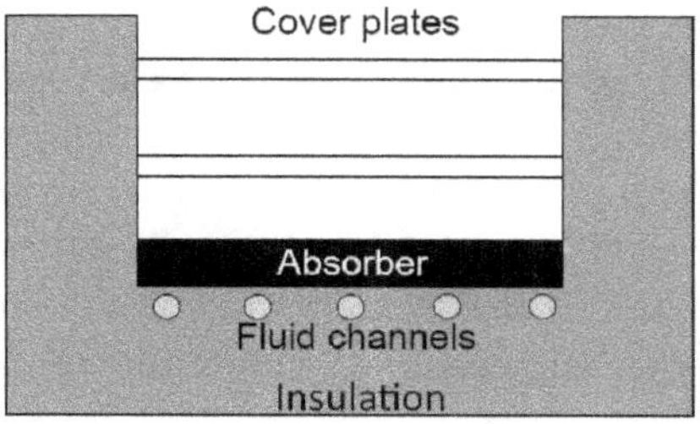

Fig. 9.2: Sketch of a solarthermal flat plate collector designed to minimize the overall heat losses.

The sunlight is incident from the top, where it first passes (typically two) transparent cover plates before hitting the absorber. These cover plates reduce heat losses due to convection. Furthermore, the collector is thermally insulated against the ambient surroundings. The losses at the back side are then due to heat conduction through the insulation and are proportional to the temperature difference between absorber and the surroundings, as evident from equation (3.18). At the front side the description of the loss mechanisms is a little more complex and can be divided into reflection and convection losses at the cover plates and radiation losses by thermal radiation from the absorber. Convection losses at the cover plates are caused by the convection of heated air on the surface of the collector and are again proportional to the temperature difference between the absorber area and the surroundings. The other two mechanisms show a different functional behavior. The reflection is independent of the absorber temperature and the radiation losses strongly depend on the material of the absorber and the cover plates and are discussed in more detail below in Section 9.1. A schematic circuit diagram for the total heat flow per unit time and unit area entering and leaving the solarthermal collector is shown in Figure 9.3.

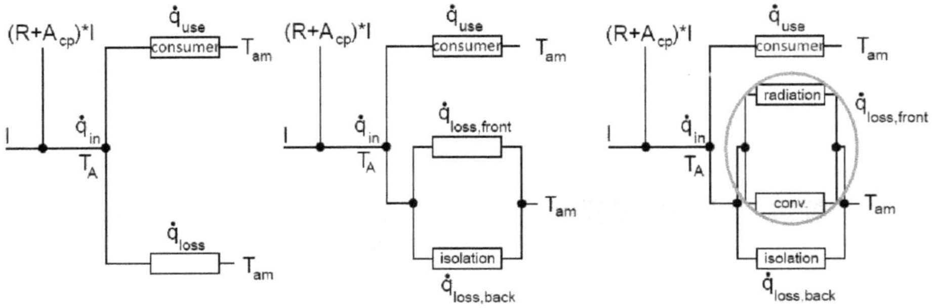

Fig. 9.3: Circuit diagrams (*left to right* with increasing details) of the heat current through a solarthermal converter. Only the heat current $\dot{q}_{\text{use}}$ is useful heat transported out of the collector by the thermal fluid, while the other pathways represent the different loss mechanisms.

The heat flow diagram in Figure 9.3 can be read in an analogous way to a conventional electric circuit, where the heat flows take the role of the electrical power and the temperature at each position that of the electrostatic potential, with the temperature $T = T_{\text{am}}$ being the electrical ground. In this picture the analog to the electrical current is the entropy flow accompanying the heat current. The input power density I is the incoming solar heat per unit time, the part $(R + A_{\text{cp}})I$ of which is reflected or absorbed by the cover plates in the first side branch on the left-hand side of each diagram. R and A_{cp} are the reflection and absorption coefficients of the stack of cover plates, respectively. Hence, the solar power $(1 - R - A_{\text{cp}})I$ reaches the absorber, where the temperature is $T = T_{\text{A}}$. The heat power incident on the absorber splits into useful heat current (upward) and losses (downward). The losses due to the imperfect insulation at the back and also the cover plates in the right-hand diagram can be regarded as being analogous to ohmic resistors, while the loss pathway denoted “radiation” has a more unusual characteristic. According to Figure 9.3, under steady-state conditions the total heat current per unit time and area through the solarthermal absorber can be written as a function of the absorber temperature T_{A} and the incident angle ϑ of the solar radiation in the following way:

$$\begin{aligned} 0 &= \dot{q}_{\text{use}}(T_{\text{A}}) + \dot{q}_{\text{loss}}(T_{\text{A}}) - I_{\perp} \cdot (1 - R(\vartheta) - A_{\text{cp}}) \\ \Rightarrow \quad \dot{q}_{\text{use}} &= I_{\perp} \cdot \left((1 - R(\vartheta) - A_{\text{cp}}) - \frac{\dot{q}_{\text{loss}}(T_{\text{A}})}{I_{\perp}} \right) \\ &=: I_{\perp} \cdot \left((1 - R(\vartheta) - A_{\text{cp}}) - \frac{k_{\text{loss}}(T_{\text{A}})}{I_{\perp}} (T_{\text{A}} - T_{\text{am}}) \right), \end{aligned} \tag{9.1}$$

where $I_\perp$ stands for the light intensity perpendicular to the collector area, and $k_{loss}(T_A)$ for the heat loss coefficient. The convection and heat-conduction losses depend linearly on the difference $T_A - T_{am}$. This is not true for radiation losses, as will be discussed in more detail below. The first-law efficiency of a solarthermal collector can then be written as

$$\begin{aligned}\eta_1 &= \frac{\dot{q}_{use}}{I_\perp} = (1 - R(\vartheta) - A_{cp})\frac{\dot{q}_{use}}{\dot{q}_{in}} \\ &=: \eta_{opt} \cdot \left(1 - \frac{k_{loss}(T_A)}{\dot{q}_{in}}(T_A - T_{am})\right),\end{aligned} \tag{9.2}$$

with the optical efficiency $\eta_{opt} = (1 - R - A_{cp})$. In the design shown in Figure 9.2 the losses due to convection at the cover plates and back-side insulation are efficiently reduced, leaving reflection and radiation as the main loss mechanisms. The reflective losses also become small compared to the radiative losses at normal incidence of the light, leaving radiation losses as the main loss mechanism.

9.1 Heat radiation of the absorber

If all loss mechanisms other than radiation losses are neglected, the absorber rejects heat by two processes only, namely the desired heating of the thermal fluid and the undesired emission of heat radiation, as shown in Figure 9.1. Even in the dark, the absorber and the thermal fluid have temperatures of $T_A = T_{TF} = T_{am}$. If the absorber is assumed to be a black body and in equilibrium with the ambient surroundings, according to the Stefan–Boltzmann law given in equation (8.7), in equilibrium with the ambient atmosphere the isolated absorber constantly emits heat radiation with the intensity $I_{am} = \sigma T_{am}^4$. Consequently, for the calculation of the total energy loss by heat radiation in the illuminated case, this intensity I_{am} has to be subtracted, as it is not caused by solar radiation. With this, in the illuminated case, the useful thermal power output via the thermal fluid $\dot{Q}_{use}$ can be calculated from a heat balance, neglecting the influence of the atmosphere on the solar radiation and again using the Stefan–Boltzmann law (8.7):

$$\frac{\dot{Q}_{use}}{A_A} = \frac{\dot{m} \cdot c_{p,TF}}{A_A} \cdot (T_{TF} - T_{am}) = \sigma\left(\frac{1}{f}T_{sun}^4 - (T_A^4 - T_{am}^4)\right), \tag{9.3}$$

where A_A is the absorber area.

In general, the expression $T_{am} \leq T_{TF} \leq T_A$ holds. For an increasing temperature T_A, the thermal radiation of the absorber increases until it emits as much radiation as it absorbs. At this point, which marks the maximal possible absorber temperature, corresponding to a flow velocity of the thermal fluid of $\dot{m} = 0$, no net energy is transferred to the thermal fluid:

$$0 = \frac{\dot{Q}_{\text{use}}}{A_{\text{A}}} = \sigma\left(\frac{1}{f}T_{\text{sun}}^4 - (T_{\text{A,max}}^4 - T_{\text{am}}^4)\right)$$
$$\Rightarrow \quad T_{\text{A,max}} = \left(\frac{1}{f}T_{\text{sun}}^4 + T_{\text{am}}^4\right)^{1/4} = 423\ \text{K} = 150\,^\circ\text{C}. \tag{9.4}$$

The useful power output can be increased by increasing the flow speed of the thermal fluid, thereby cooling the absorber and reducing its temperature. The power output is maximized when $T_{\text{A}} \to T_{\text{am}}$, where it approaches the full power of the solar radiation as the difference in the heat radiation from the absorber in the illuminated and the unilluminated case vanishes. At the same time, in this limit the value of the harvested energy, i.e. the area specific exergy per unit time transferred to the thermal fluid $\dot{\Psi}_{\text{use}}/A_{\text{A}}$, decreases to zero as pure anergy is then transferred.

The exergy input per unit area and time is a function of both the energy input and the absorber temperature. For a given absorber temperature T_{A} it is given by

$$\dot{\Psi}_{\text{use}} = \dot{Q}_{\text{use}}\left(1 - \frac{T_{\text{am}}}{T_{\text{A}}}\right) = \dot{m} \cdot c_{\text{p,TF}}(T_{\text{A}}) \cdot (T_{\text{A}} - T_{\text{am}})\left(1 - \frac{T_{\text{am}}}{T_{\text{A}}}\right). \tag{9.5}$$

This exergy decreases even further if the difference between absorber temperature T_{A} and temperature of the thermal fluid T_{TF} necessary to drive the heat transfer from the former to the latter is taken into account. The maximal transferable energies and exergies per unit area and unit time $\dot{Q}/A_{\text{A}}$ and $\dot{\Psi}/A_{\text{A}}$ transferred to the thermal fluid according to equations (9.3) and (9.5) are shown in Figure 9.4 as a function of the absorber temperature.

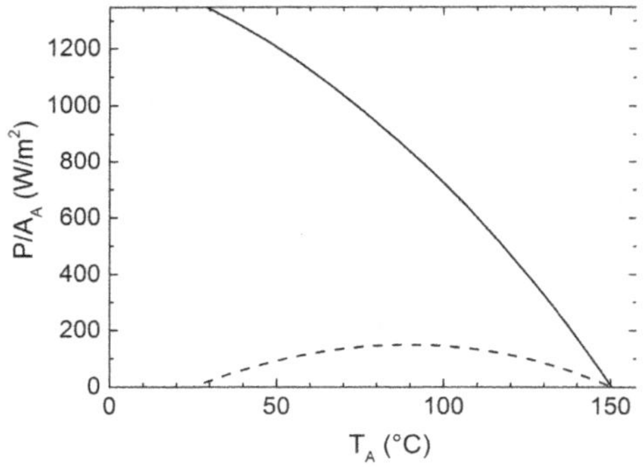

Fig. 9.4: Maximal energy per unit absorber area (*solid curve*) and exergy per unit absorber area (*dashed curve*) transferred to the thermal fluid of a solarthermal collector with only heat radiation as a loss mechanism.

It is apparent from the exergy transfer curve in Figure 9.4 that solarthermal flat plate collectors without further modifications are not useful in the production of electrical energy. As the exergy content of the heat transferred at low temperatures for room heating and warm water generation in a typical household is very small, however, solarthermal flat plate collectors are still useful for providing this low temperature heat in an efficient way.

9.1.1 Selective absorption

To increase the first- and second-law efficiencies of solarthermal energy collectors the heat radiation from the absorber has to be reduced. If a piece of matter has identical absorption and emission coefficients equal to 1 for all photon energies, it is a black body. If, on the other hand, the absorption coefficient A is different from 1 in any given photon energy interval $[E, E + dE]$, the nonabsorbed part of the light is transmitted, which together with the reflection at the surface leads to the following expression linking the absorption, reflection and transmission coefficients:

$$1 = A(E) + T(E) + R(E). \tag{9.6}$$

The spectral energy distribution of the solar photons is determined by the temperature of the solar surface, while the spectral distribution of heat radiation from the absorber is determined by the absorber temperature. This means that the solar radiation delivers most of its energy in the near-infrared and the visible (0.5–3 eV) spectral range, while the heat emission from an absorber at e.g. 100 °C peaks at 0.09 eV, as shown in Figure 9.5.

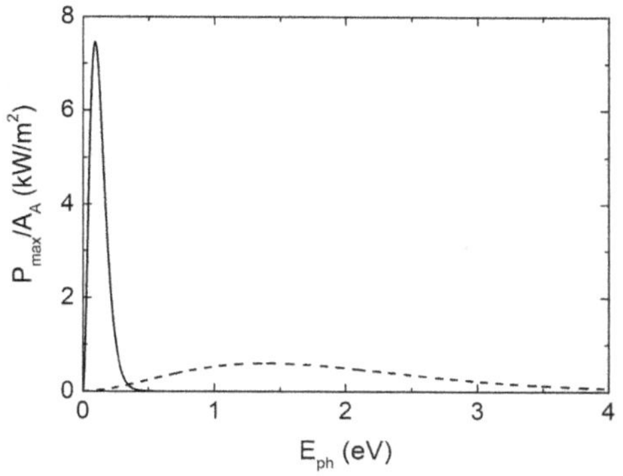

Fig. 9.5: Black body absorber heat radiation emission spectrum at $T_A = 100$ °C (*solid curve*), and the incoming solar radiation spectrum (*dashed curve*).

Due to the small overlap of both spectra it is possible to increase the overall efficiency of a solarthermal collector if it absorbs well in the region of comparatively high photon energies (0.5–3 eV) while absorbing and thus emitting poorly in the region of photon energies below this energy interval. The former assures a good absorption of the solar irradiation, and the latter suppresses the thermal radiation of the absorber. For a given absorber temperature the losses of heat radiation from the emitter can thereby be dramatically reduced. The proposed absorber modification is called **selective absorption** and can in general be achieved in two qualitatively different ways. First, it is possible to choose the absorber material itself in a way that it shows an absorption characteristic favoring high photon energies and, second, the cover plates can be coated with a coating having a spectrally selective transmission coefficient. In Figure 9.6 the effect of a cut-off selective absorber, i.e. an absorber that only absorbs photons with energies above a threshold $E_{ph} > E_{th}$ is shown for a threshold energy of $E_{th} = 0.5$ eV.

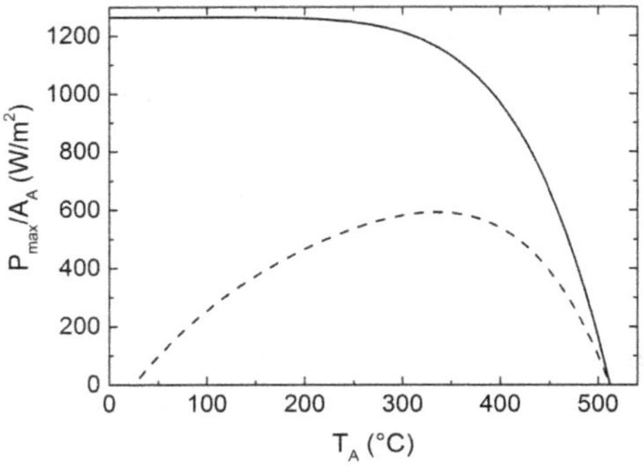

Fig. 9.6: Maximal energy per unit absorber area (*solid curve*) and exergy per unit absorber area (*dashed curve*) transferred to the thermal fluid of a solarthermal collector with an absorber selectively absorbing above a threshold of $E_{th} = 0.5$ eV. Heat radiation is assumed to be the only loss mechanism.

A comparison between Figures 9.4 and 9.6 reveals the pronounced positive effect of the selective absorption to the overall first- and second-law efficiencies for a wide temperature range and also shows that much higher temperatures can be achieved. Probably the most famous example of selective absorption is the greenhouse effect caused by the atmosphere of the earth, as described in Section 8.2.1. As already indicated in the name, the atmosphere of the earth can be seen as analogous to a selectively transmitting glass which suppresses the heat emission from the surface of the earth while having a smaller effect on the incoming solar radiation.

9.2 Concentrating devices

Concentration of the solar irradiation often becomes necessary, because the intensity of the sunlight at the surface of the earth is not high enough for many solar energy applications, most prominently for solarthermal power plants. For these applications temperatures higher than those achievable with unconcentrated light are necessary.

9.2.1 Imaging optics

An imaging system produces an image of the visible solar surface on the absorber. A criterion for the production of a sharp image from rays close to the optical axis of any imaging system is the **Abbe sine condition**:

$$\sin(\alpha) = \frac{1}{C} \cdot \sin(\beta). \tag{9.7}$$

Here α stands for the incident angle of the rays coming from the object, β for the incident angle of the rays from the image, as shown in Figure 9.7, and C for the concentration factor, i.e. the ratio between the object and the image size.

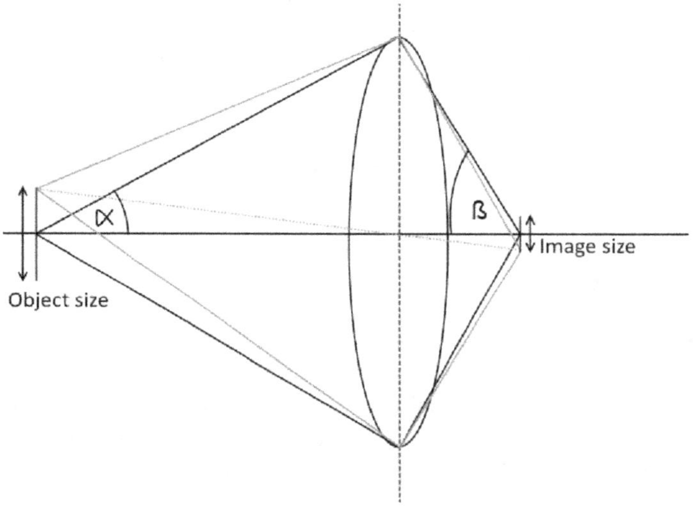

Fig. 9.7: Production of a sharp image (*right*) of an object (left) through an imaging system (*oval*) as given by the Abbe sine condition. All rays from every point at the object are incident to the imaging system with angles smaller than α and all rays from the imaging system to the image are leaving on the imaging system with angles smaller than β.

When the light source is much larger than the inlet aperture of the imaging system, as is the case for the sun, the image generated is an image of the illuminated aperture, as the image is only constituted by rays passing through the aperture. Rays originating from every point of the aperture have a divergence α that corresponds to the opening angle of the sun. It can be calculated using the solar radius r_s and the distance d between sun and earth:

$$\alpha = \arctan\left(\frac{r_s}{d}\right) \approx \frac{r_s}{d} = 4.65 \cdot 10^{-3} = 0.267^\circ. \tag{9.8}$$

The Abbe sine condition is true for both directions perpendicular to the principal direction of the incident rays, which leads to the following expression for an undistorted image via 2d focusing:

$$\sin^2(\alpha) = \frac{1}{C} \cdot \sin^2(\beta). \tag{9.9}$$

As $|\sin(\beta)| \leq 1$, it is possible to calculate the maximal concentration factor C_{max} achievable for 1d and 2d focusing with an imaging device:

$$\begin{aligned} C_{\mathrm{max}} &= \frac{1}{\sin^d(\alpha)}; \qquad d \in \{1,2\} \\ \Rightarrow \quad C_{\mathrm{max,1d}} &= 212 \\ \Rightarrow \quad C_{\mathrm{max,2d}} &= 46200 = f. \end{aligned} \tag{9.10}$$

A very interesting consequence of the Abbe sine condition arises when the maximal focusing possible according to equation (9.10) is related to the Stefan–Boltzmann law, as given in equation (8.7). The temperature of the absorber with area A_{A} under maximal 2d focusing according to the Stefan–Boltzmann law is

$$\begin{aligned} \sigma T_{\mathrm{A}}^4 \cdot A_{\mathrm{A}} &= \frac{1}{f}\sigma T_{\mathrm{s}}^4 \cdot A_{\mathrm{A}} \cdot C_{\mathrm{max,2d}} \\ \Rightarrow \quad T_{\mathrm{A}}^4 &= \frac{C_{\mathrm{max,2d}}}{f} T_{\mathrm{s}}^4 = T_{\mathrm{s}}^4 \\ \Rightarrow \quad T_{\mathrm{A}} &= T_{\mathrm{s}}. \end{aligned} \tag{9.11}$$

This means that under the maximal focusing possible given by the Abbe sine condition, the temperature of the absorber T_{A} reaches the temperature of the solar surface T_{s}. In other words, the absorber and the solar surface are in thermal equilibrium. The value for the maximal absorber temperature T_{A} found by using the Abbe sine condition is exactly the upper limit for T_{A} given by the second law of thermodynamics in its formulation by Clausius, as laid out in Section 2.1,

$$T_{\mathrm{A}} \leq T_{\mathrm{s}}, \tag{9.12}$$

as heat can only be transferred spontaneously from a source at higher temperature to a target with lower temperature. This consideration can also be used as the starting point for the derivation of the maximal possible focusing.

When the size of the inlet aperture A_{Ap} of the imaging system is increased, a growing part of the diffuse solar radiation is imaged on the absorber. The opening angle of the object imaged is then no longer the opening angle of the solar disk visible from the surface of the earth, but also an increasing disk of the surrounding sky is imaged. This means that the angle α is increased, which directly implies that the maximal possible concentration factor decreases with increasing aperture size, as can be seen from equation (9.10). The maximal absorber temperature achievable is thus reduced. In elevated latitudes, as for example in central Europe, only about 40 % of the solar radiation is direct and focusing systems are not particularly usable there. They show the best performance in regions closer to the equator, for example in the Sahara desert or the south of the US, where 70–80 % of the solar radiation is direct radiation.

9.2.2 Technical realization

Focusing is typically achieved by concave mirrors, as shown in Figure 9.8, and can be classified into systems with a rotational symmetry (dishes or fields of heliostats) for 2d focusing and systems with a translational symmetry (troughs) for 1d focusing.

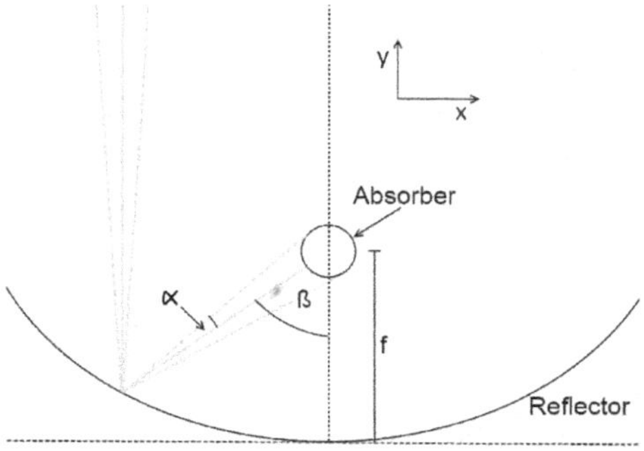

Fig. 9.8: Cut through a concave mirror with the focal length f for the focusing of solar radiation onto an absorber. The angles α and β are relevant for the Abbe sine condition equation (9.7).

For both 1d and 2d focusing, the optimal focusing can be achieved by a parabolic form of the focusing mirror, satisfying the following condition:

$$y = \frac{x^2}{4f}, \tag{9.13}$$

where y and x are the distances to the vertex of the parabola parallel and perpendicular to the incident radiation, respectively, and f is the focal length. For technical reasons it is often easier to produce concave mirrors as segments of a circle rather than parabolas. For such a segment of a circle with radius R, the curve (x, y) of all points on the segment with the vertex in the origin is given by

$$x^2 + (y - R)^2 = R^2 \underset{y<R}{\Rightarrow} y(x) = R - R \cdot \sqrt{1 - \frac{x^2}{R^2}}. \tag{9.14}$$

The focal length of such a mirror is $f = R/2$. If only a small segment is taken, i.e. if $x \ll R$, equation (9.14) can be approximated by expanding it into a Taylor series in x^2/R^2 up to the linear term:

$$y(x) \approx \frac{x^2}{2R}. \tag{9.15}$$

Substituting $f = R/2$, this yields exactly the result given in equation (9.13). The maximal 1-dimensional focusing that can be achieved by such a segment of a circle can be calculated using equation (9.7), taking into account that the absorber area completely covers the circle denoted as "Absorber" in Figure 9.8 and not only the area perpendicular to the incoming radiation, which leads to an additional factor $1/\pi$:

$$C = \frac{1}{\pi}\frac{\sin(\beta)}{\sin(\alpha)} = \frac{\sin(\beta)}{\pi} C_{\text{max,1d}}. \tag{9.16}$$

9.2.3 Efficiency of concentrating solarthermal energy converters

In a realistic device a number of loss mechanisms can occur. These loss mechanisms are partly identical to the ones discussed for the flat plate collector in the beginning of this chapter, and partly imperfections of the mirrors and geographical effects which worsen the image quality and hence the achievable focusing. The geographical effects are omitted here but need to be considered when a location for a focusing solarthermal converter is chosen, because they limit the performance of the converter as a function of, e.g. the incident angle of the solar radiation ϑ. The heat current balance is similar to the one for flat plate collectors given in equation (9.1). Only the details change: for example the loss due to imperfect reflection of the mirrors has to be taken into account because the mirror is an additional element in the light path. Again the convection and heat conduction losses can be efficiently reduced by an appropriate thermal insulation. As already laid out in Section 9.1, the only loss mechanism which cannot be completely suppressed due to thermodynamics is the thermal heat radiation from the absorber. Following the argumentation from Section 9.1 as laid out in equations (9.3)–(9.5), the extractable exergy per unit time and unit aperture area $\dot{\psi}_{\mathrm{out}}$ as a function of the absorber temperature in an isothermal process for maximal 1d and 2d focusing can be expressed as follows:

$$\begin{aligned} \text{1d:} \quad \dot{\psi}_{\mathrm{out}} &= \frac{\sigma}{\sqrt{f}} \left(\frac{1}{\sqrt{f}} T_{\mathrm{sun}}^4 - (T_{\mathrm{A}}^4 - T_{\mathrm{am}}^4) \right) \cdot \left(1 - \frac{T_{\mathrm{am}}}{T_{\mathrm{A}}} \right) \\ \text{2d:} \quad \dot{\psi}_{\mathrm{out}} &= \frac{\sigma}{f} \left(T_{\mathrm{sun}}^4 - (T_{\mathrm{A}}^4 - T_{\mathrm{am}}^4) \right) \cdot \left(1 - \frac{T_{\mathrm{am}}}{T_{\mathrm{A}}} \right) . \end{aligned} \tag{9.17}$$

The corresponding second-law efficiencies can be calculated using the total exergy input from the sun, as given in equation (8.11). They are shown in Figure 9.9, with the maxima of the curves being at $T_{\mathrm{A;1d}} = 883\ \mathrm{K} = 610\ ^{\circ}\mathrm{C}$, with $\eta_2 = 66\ \%$ for 1d focusing and $T_{\mathrm{A;2d}} = 2473\ \mathrm{K} = 2200\ ^{\circ}\mathrm{C}$ with $\eta_2 = 93\ \%$ for 2d focusing.

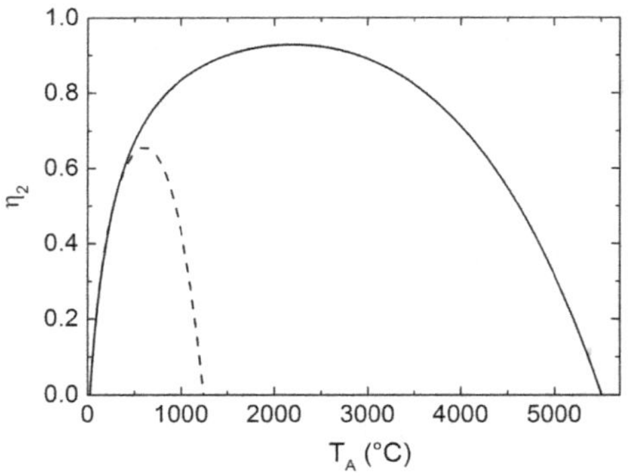

Fig. 9.9: Maximal second-law efficiency of an isothermal solarthermal energy converter as a function of the absorber temperature for a converter with 1d focusing (*dashed curve*) and 2d focusing (*solid curve*).

Again, the radiative losses can be decreased by the use of selective absorbers. The effect on the extractable exergy is, however, less pronounced than for the flat plate collectors in Section 9.1.1 for increasing absorber temperatures as the spectral mismatch between incoming and emitted radiation is smaller. Still, at least for 1d focusing selective absorption increases the performance of the solarthermal converter. In real solarthermal converters the heat transfer does not occur isothermally, because the thermal fluid is passing the absorber, thereby increasing its temperature from $T = T_c$ to $T = T_h$ in an isobaric process. Even under the assumption of a constant heat capacity $c_p = q/(T_h - T_c)$ of the thermal fluid, the total energy and exergy transfers into the thermal fluid cannot be analytically calculated, because the absorber itself has a nonconstant temperature and the heat transfer a nonlinear temperature dependence. Numerical calculations of the second-law efficiencies of the isobaric heat transfer process for 1d and 2d focusing are shown in Figure 9.10.

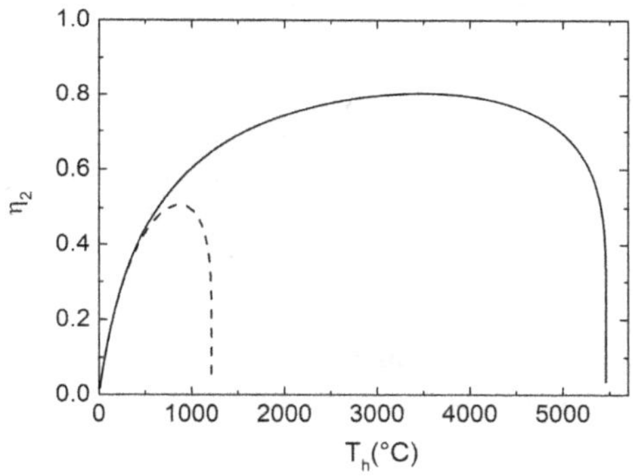

Fig. 9.10: Maximal second-law efficiency of an isobaric solarthermal energy converter as a function of the maximal absorber temperature T_h with $T_c = T_{am}$ for a converter with 1d focusing (*dashed curve*) and 2d focusing (*solid curve*).

The maximal second-law efficiencies are shifted towards higher temperatures in this case. They are at $T_{h,2d} = 3753$ K $= 3480$ °C with $\eta_2 = 80$ % for 2d focusing and at $T_{h,1d} = 1148$ K $= 875$ °C with $\eta_2 = 50$ % for 1d focusing.

9.2.4 Solarthermal power plants

In areas where a large fraction of the solar radiation is direct, concentrating solarthermal power plants can be operated. Most solarthermal power plants are steam power plants, where the working fluid water is heated indirectly by solar radiation. A big advantage of solarthermal power plants as opposed to photovoltaic fields is that heat storage tanks, typically molten-salt heat capacity tanks (which will be discussed in Section 11.2.1) can be used to store a part of the collected solar energy. In this way a continuous operation can even be achieved at night. A sketch of the working principle of a typical solarthermal power plant is shown in Figure 9.11.

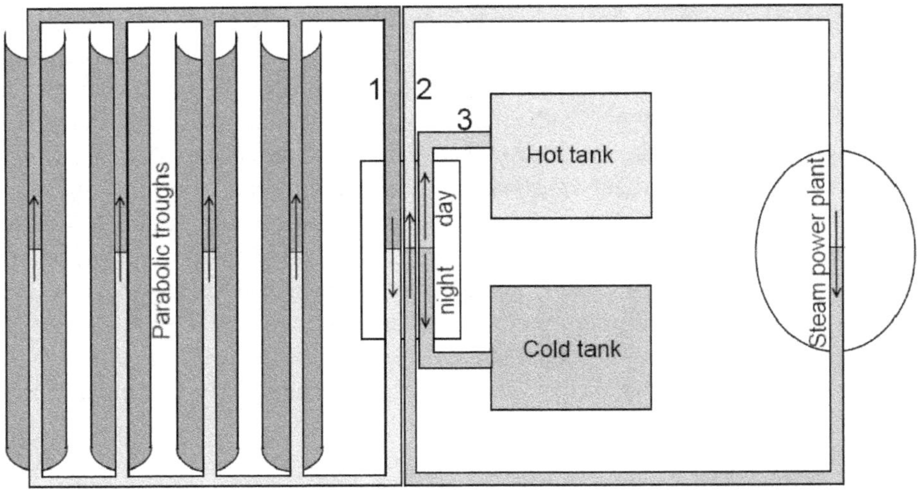

Fig. 9.11: Schematic of a solarthermal parabolic trough power plant. The thermal oil is heated in the collector field during the day and circles through the tubing system 1 in the direction indicated, heating the working fluid of the steam power plant (tubing system 2) and molten salt pumped from the low temperature storage to the high temperature storage in tubing system 3. During the night the flow direction in tubing system 3 is inverted, thus heating the working fluid of the steam power plant with the molten salt.

Solarthermal power plants can be classified into **solar trough power plants**, which use 1d focusing (see Figure 9.11), and **solar tower power plants**, which use 2d focusing. In the former type, a field of parabolic trough collectors is used, where the 1d focal lines are occupied by the absorber tubes carrying the thermal oil as part of a tubing system. The temperature in such plants typically reaches values of ca. $T_{max} \approx 400\,^{\circ}\mathrm{C}$. The troughs are oriented along the north–south axis, which makes an east–west tracing of the sun possible by tilting the troughs. In the latter type a field of mirrors (heliostats) constituting a **Fresnel lens** are used to focus the solar radiation onto a single absorber area located in the top section of a tower. With this design a tracing of the sun along both the east–west and up–down directions is possible. Advantages compared to the solar trough power plants are, apart from the better tracing of the sun, the higher temperatures possible because of the 2d focusing and the lack of extended tubing systems carrying heated fluid. These advantages, however, are offset by the difficulties in handling the high temperatures of the absorber, which is why both technologies are currently competitive. Current solarthermal power plants reach first law efficiencies of ca. $\eta_1 = 15\,\%$.

10 Photovoltaic energy conversion

In a photovoltaic device, solar energy is converted into electricity along a path very different from the one taken in a solarthermal power plant. Here, in a first step the energy of the solar photons is converted into chemical energy in a solid state absorber. This means that the absorber is brought into an electronically excited state involving a reconfiguration of its charge carriers by the generation of electron/hole (e^-/h^+)-pairs, i.e. by the following reaction:

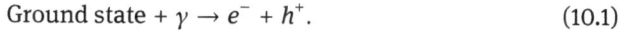

$$\text{Ground state} + \gamma \rightarrow e^- + h^+. \tag{10.1}$$

Here, γ represents a photon with sufficient energy to bring an electron to the excited state. The chemical energy of the charge carrier ensembles in the conduction and valence bands is then converted into electrical energy by spatially separating the e^-/h^+-pairs via electrical contacts of the absorber which are electron or hole selective, respectively. In general such selective contacts can only be realized by a jump in the material properties between the two contacts, an example for this being a pn-junction. Since under illumination electrons and holes have different electrochemical potentials in the absorber material, this separation leads to a voltage drop between the contacts selective for the different charge carrier types. It is thus the selectivity of the contacts that introduces the built-in asymmetry into the solar cell, making it a usable voltage source (see Section 5.2). This basic working principle is true for all types of solar cells, ranging from conventional solar cells built from crystalline silicon (c-Si) over thin film solar cells fabricated from different materials such as, e.g. Cu(In,Ga)Se_2 (CIGS) to organic or dye sensitized solar cells, and is schematically shown in Figure 10.1.

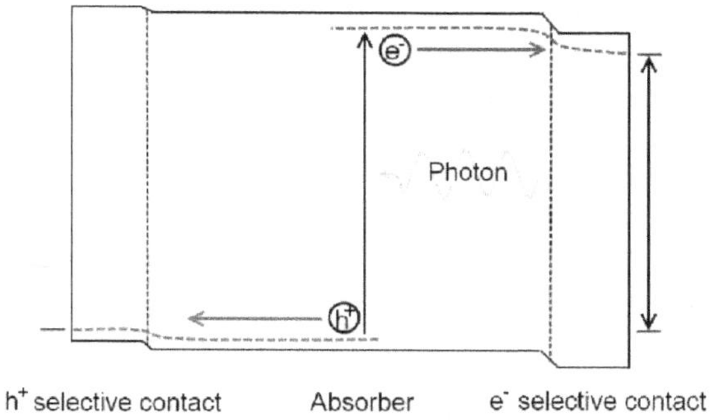

Fig. 10.1: Working principle of a photovoltaic cell.

In general a solar cell is built such that every photon with an energy larger than the bandgap, i.e. $E_{ph} > E_g$, generates an e^-/h^+-pair, and thus the total generation rate of these pairs is constant for a given intensity of the incoming irradiation. These photogenerated mobile charge carriers then leave the absorber volume in one of two ways. First, they can contribute to the external current I, i.e. they leave the absorber through

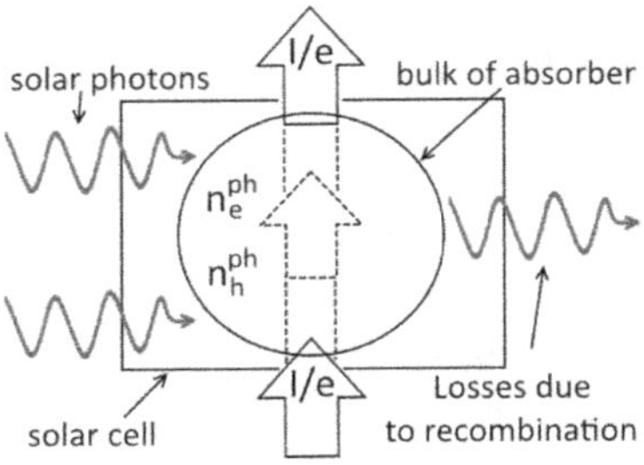

Fig. 10.2: Schematic of a photovoltaic cell as an open system. The flow rate is given by the external current, and the energy gain of the working medium when crossing the open system is determined by the densities of photogenerated mobile charge carriers in the bulk of the absorber, n_e^{ph} and n_h^{ph}.

the electrical contacts. This is the desired mechanism, as the energy and exergy of them can then be externally used. Second, the photogenerated electrons and holes are not stable and recombine after a certain lifetime, which in the case of crystalline silicon is in the order of ≈ 100 μs. In this recombination process heat is released, either in the form of phonons or as heat radiation, and the energy is no longer externally usable. In operation, a solar cell can be seen as an open system, with the ensemble of photoexcited charge carriers as the working fluid. At a given irradiation and a defined external load this open system is then in a steady flow equilibrium, with the particle flow I/e being the flow rate, as shown in Figure 10.2. Note that each e^-/h^+-pair only counts as one charge carrier for the current. The reason for this can be seen in Figure 10.1, where the electron and the hole each carry the current only a part of the total distance through the solar cell. The question is then how much energy and exergy an electron gains when passing through the solar cell, i.e. entering it at the hole selective contact annihilating a hole and leaving it at the electron selective contact. This energy and exergy gains are linked to the density of photogenerated electrons n_e^{ph} and holes $n_h^{ph} = n_e^{ph}$ in the bulk of the absorber. Summing this up, the steady flow equilibrium is determined by the irradiation and the external load and characterized by the flow rate, i.e. the current, and the density of photogenerated charge carriers, which in turn determines the energy and exergy gain of a charge carrier crossing the system. This is schematically shown in Figure 10.2, where the rectangle represents the entire solar cell and the circle the bulk of the absorber outside the selective contacts.

The energy and exergy gains of a charge carrier passing the photovoltaic open system are chemical energy, as they are linked to the insertion and extraction of particles and thus analogous to a chemical reaction. Both are linked to the change of the Gibbs free energy with the number of particles, i.e. the chemical energy of the particles.

In the following sections the physical principles of the conversion of solar energy into chemical energy and of chemical energy into electrical energy will be discussed in detail for c-Si solar cells. First the relationship between the photogenerated charge carrier densities n_e^{ph} and n_h^{ph} in the bulk of the absorber and its chemical energy and exergy will be derived. This will be followed by a discussion of the interplay between generation, extraction, and recombination of photogenerated charge carriers determining the steady flow equilibrium, which emerges in the system due to the illumination.

10.1 Conversion of solar into chemical energy

For the first energy conversion step, the bulk absorber material and its properties upon excitation by the solar photons are considered. Here, only the bulk absorber is of interest, and any surface effects, especially those associated with the selective contacts, are neglected. In a c-Si solar cell the absorber is lowly p-type doped silicon of several hundred micrometers thickness. The energy distribution of the holes in the valence band in the dark $dn_\mathrm{h}^\mathrm{dark}/dE$ is discussed in the appendix and shown in Figure B.2. Upon photoexcitation, first electrons from the valence band are excited into the conduction band, leaving behind a corresponding amount of holes. The initial energetic distribution of these photoexcited electrons and holes is highly unstable, and interactions with the silicon crystal lattice lead to an immediate (≈ 1 ps) relaxation into a more stable configuration in a process called **thermalization**. In this process, the photogenerated charge carriers attain thermal equilibrium with the crystal lattice. However, the number of photoexcited charge carriers in the respective bands is not changed by the thermalization process. They merely change their energetic distribution. It is this energetic distribution of the photogenerated charge carriers after thermalization that is of interest, as the timescale of the thermalization process is too short to extract, rather than lose, the energy in any realistic device. Only on a much longer timescale in the order of ≈ 100 μs do the photogenerated charge carriers generated in both bands annihilate by recombination, a timescale where extraction is quite feasible.

10.1.1 Chemical energy and exergy at a given n_e and n_h

In a given steady-flow equilibrium a certain charge carrier density n_e and n_h is present in both bands. Due to the typical p-type doping of the absorber material the following expressions hold:

$$\begin{aligned} n_\mathrm{e} &= n_\mathrm{e}^\mathrm{ph} + n_\mathrm{e}^\mathrm{dark} \approx n_\mathrm{e}^\mathrm{ph} \\ n_\mathrm{h} &= n_\mathrm{h}^\mathrm{ph} + n_\mathrm{h}^\mathrm{dark} > n_\mathrm{h}^\mathrm{ph}, \end{aligned} \tag{10.2}$$

where the index 'ph' indicates the photogenerated charge carriers. As electrons and holes are created pairwise, both photoinduced contributions are identical: $n_\mathrm{e}^\mathrm{ph} = n_\mathrm{h}^\mathrm{ph}$. These charge carriers are already thermalized, i.e. they are in a relatively long-lived configuration which is in thermal equilibrium with the silicon crystal lattice. Both holes and electrons are fermions, and thus the energetic spectra dn_e/dE and dn_h/dE of these charge carrier ensembles obey the following relations:

$$\begin{aligned} dn_\mathrm{e}/dE &= D_\mathrm{CB}(E) f_\mathrm{CB}(E,T) = D_\mathrm{CB}(E)\frac{1}{e^{(E-\tilde{\mu}_\mathrm{CB})/kT}+1} \\ dn_\mathrm{h}/dE &= D_\mathrm{VB}(E)(1-f_\mathrm{VB}(E,T)) = D_\mathrm{VB}(E)\frac{1}{e^{(\tilde{\mu}_\mathrm{VB}-E)/kT}+1}. \end{aligned} \tag{10.3}$$

Here $D_{CB,VB}(E)$ stands for the density of electronic states at a given energy E in the conduction and valence band, respectively, and $f_{CB,VB}(E,T)$ is the Fermi–Dirac distribution determining the probability that a state at energy E is occupied by an electron in the respective bands. The electrochemical potentials of the electrons in the conduction band and the holes in the valence band are denoted by $\tilde{\mu}_{CB}$ and $\tilde{\mu}_{VB}$, respectively. A more detailed derivation of all relevant concepts and quantities in equation (10.3) can be found in Appendix B. For a visualization the energy spectrum of the electrons in the conduction band is shown together with the density of states and the Fermi–Dirac distribution function in Figure 10.3.

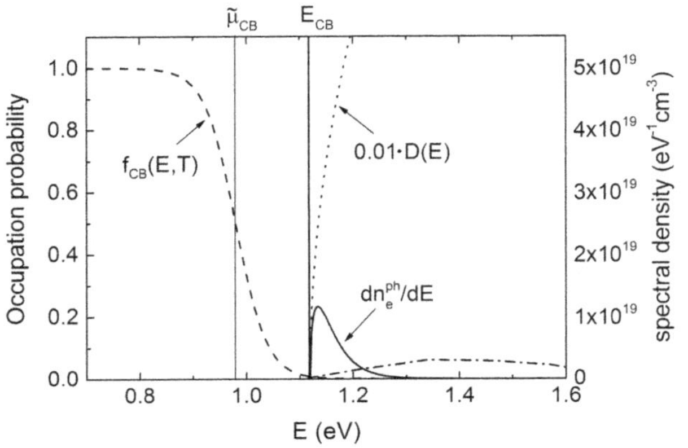

Fig. 10.3: Energetic distribution of the photogenerated electrons in the conduction band dn_e^{ph}/dE (*solid curve, right scale*) of illuminated p-doped silicon at $T = 60\,^\circ\mathrm{C}$ together with the Fermi–Dirac distribution $f_{CB}(E,T)$ (*dashed curve, left scale*) and density of states (*dotted curve, right scale*) of the conduction band. The energy distribution before thermalization is also qualitatively indicated (*dash-dot curve*).

Using equation (10.3), the photoinduced electron and hole densities can be calculated by integration over the respective bands. Using the approximation $|\tilde{\mu} - E| > 3kT$, which is typically valid for all relevant energies in both bands, the following expressions are found:

$$\begin{aligned} n_e &= \int_{E_{CB}}^{\infty} D_{CB}(E) f_{CB}(E,T)\,dE =: N_C \cdot e^{-(E_{CB}-\tilde{\mu}_{CB})/kT} \\ n_h &= \int_{-\infty}^{E_{VB}} D_{VB}(E)(1 - f_{VB}(E,T))\,dE =: N_V \cdot e^{-(\tilde{\mu}_{VB}-E_{VB})/kT}, \end{aligned} \tag{10.4}$$

where $E_{CB,VB}$ stands for the conduction and valence band edge, respectively, and N_C and N_V represent the so-called **effective density of states** in both bands. These latter quantities are material constants and are of the order of 10^{19} cm^{-3} for silicon. A more detailed discussion of the integral in equation (10.4) can be found in Section B.2.1 of the appendix. The only free parameters in equation (10.4) are the two electrochemical potentials $\tilde{\mu}_{CB}$ and $\tilde{\mu}_{VB}$. These electrochemical potentials are not identical under illumination. $\tilde{\mu}_{CB}$ lies relatively close to the conduction band edge while $\tilde{\mu}_{VB}$ lies relatively close to the edge of the valence band. Since the occurrence of two distinct electrochemical potentials does not correspond to the equilibrium state where only one electrochemical potential is found throughout the system, $\tilde{\mu}_{CB}$ and $\tilde{\mu}_{VB}$ are called **quasi-electrochemical potentials.** To emphasize this again keep the following in mind.

Under illumination the charge carrier ensembles in valence and conduction band are not in their equilibrium state. After thermalization in their respective bands they obey separate Fermi–Dirac distributions, and the corresponding electrochemical potentials are called quasi-electrochemical potentials $\tilde{\mu}_{VB}$ and $\tilde{\mu}_{CB}$.

As a last note on this, in the same way that the electrochemical potential is often denoted as “Fermi energy”, the quasi-electrochemical potentials are typically called “quasi Fermi levels” or “quasi Fermi energies”. In the opinion of the authors this obscures the meaning and seems to be related to the very common but slightly incorrect use of “Fermi energy” instead of “electrochemical potential” in other contexts.

The next step is the calculation of chemical energy and exergy of the two charge carrier ensembles. For the average energy difference per photoinduced particle to the respective band edges of the two charge carrier ensembles in valence and conduction band $\varepsilon_{av,e/h}$ the following expression can be found (see also equation (B.10) in the appendix):

$$\begin{aligned} \varepsilon_{av,e} &:= \frac{1}{n_e} \int_{E_{CB}}^{\infty} (E - E_{CB}) \cdot D_{CB}(E) \cdot f_{CB}(E, T) dE \\ \varepsilon_{av,h} &:= \frac{1}{n_h} \int_{-\infty}^{E_{VB}} (E_{VB} - E) \cdot D_{VB}(E) \cdot (1 - f_{VB}(E, T)) dE, \end{aligned} \tag{10.5}$$

where $D_{CB,VB}$ are the densities of states and $f_{CB,VB}$ the Fermi–Dirac distribution functions in the conduction and valence band, respectively. Again using the approximation $|\tilde{\mu} - E| > 3kT$ for all energies in the bands and the expression

$$\int_0^{\infty} x^{3/2} \cdot e^{-x} dx = \frac{3}{4}\sqrt{\pi}, \tag{10.6}$$

the integrals in equation (10.5) can be calculated. This calculation is completely analogous to equation (B.23) in the appendix, and the following average energy distance to the band edge per electron $\varepsilon_{av,e}$ in the conduction band is found:

$$\varepsilon_{av,e} = \frac{3/2kT \cdot N_C e^{-(E_{CB}-\tilde{\mu}_{CB})}}{N_C e^{-(E_{CB}-\tilde{\mu}_{CB})}} = \frac{3}{2}kT. \tag{10.7}$$

The same value is obtained for the average energy distance of the photoinduced holes to the valence band edge $\varepsilon_{av,h} = 3/2kT$. If the average energy difference per charge carrier to the respective band edge is interpreted as its kinetic energy, it becomes evident from equation (10.7) that the electrons in the conduction band and the holes in the valence band can be seen as an ideal gas, provided that their density is not to high, i.e. within the approximation $|\tilde{\mu} - E| > 3kT$. The average chemical energy which can be released by each e^-/h^+-pair recombination, i.e. by the transfer of an excited electron in the conduction band to the unoccupied state in the valence band represented by the hole, is then consequently given by

$$\Delta H_{rec} = -(E_{CB} + \varepsilon_{av,e} - E_{VB} - \varepsilon_{av,h}) = -(E_g + 3kT). \tag{10.8}$$

With this, the extractable photoinduced chemical energy per unit volume $-\Delta h_{rec}$ can be given as a function of the photoinduced charge carrier density, which for p-doped material is identical to the electron density $n_h^{ph} = n_e^{ph} = n_e$:

$$-\Delta h_{rec} = -n_e^{ph}\Delta H_{rec} = n_e \Delta H_{rec} = n_e(E_g + 3kT). \tag{10.9}$$

The disordered movement of the charge carriers in both conduction and valence band leads to an entropic component of the internal energy of the excited absorber state. The exergy gain Ψ of an electron crossing the photovoltaic cell is given by the difference in the electrochemical potential of the electron entering the cell at the hole selective contact, annihilating a hole, and the electron leaving it at the electron selective contact at the same time. As already shown in Chapter 5 the exergy gain is thus simply given by the difference in the electrochemical potentials of both interfaces:

$$\Psi = \tilde{\mu}_{CB} - \tilde{\mu}_{VB}. \tag{10.10}$$

This leads to a photoinduced chemical exergy per unit volume ψ:

$$\psi = n_e^{ph}\Psi = n_e\Psi = n_e \cdot (\tilde{\mu}_{CB} - \tilde{\mu}_{VB}). \tag{10.11}$$

The chemical exergy per unit volume is thus only determined by the quasi-electrochemical potentials. In turn, the positions of the quasi-electrochemical potentials with respect to the band edges $E_{CB} - \tilde{\mu}_{CB}$ and $\tilde{\mu}_{VB} - E_{VB}$ for the given electron and hole densities n_e and n_h can be obtained using equation (10.4):

$$\begin{aligned} E_{CB} - \tilde{\mu}_{CB} &= kT \cdot ln\left(\frac{N_c}{n_e}\right) \\ \tilde{\mu}_{VB} - E_{VB} &= kT \cdot ln\left(\frac{N_v}{n_h}\right). \end{aligned} \tag{10.12}$$

Substituting equation (10.12) into equation (10.11), this leads to the following value for the photoinduced chemical exergy per unit volume ψ:

$$\psi = n_e \cdot (\tilde{\mu}_{CB} - \tilde{\mu}_{VB}) = n_e \left(E_g - kT \cdot \ln \left(\frac{N_C N_V}{n_e n_h} \right) \right). \tag{10.13}$$

This means that the chemical exergy per unit volume ψ produced in the system by the solar irradiation decreases linearly with temperature and increases with the charge carrier density imposed by the solar radiation. In Figure 10.4 the photoinduced chemical energy $-\Delta H_{rec}$ and exergy Ψ per particle are qualitatively shown.

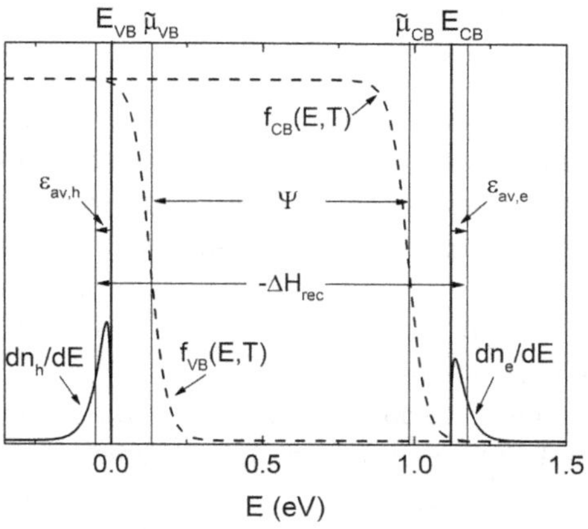

Fig. 10.4: Average energy $-\Delta H_{rec}$ and exergy Ψ per e^-/h^+-pair together with the energetic distribution of holes in the valence band dn_h/dE and electrons in the conduction band dn_e/dE of illuminated p-doped silicon at $T = 60\,^\circ\mathrm{C}$ (*solid curves*) and the Fermi-Dirac distributions (*dashed curves*) for valence and conduction band.

10.1.2 Radiative recombination

For further considerations, a closer look at the interplay between the volume-specific rates of photoinduced electron and hole generation r_{gen}, recombination r_{rec}, and external current has to be taken. It is this interplay that determines the steady flow equilibrium which was so far just assumed as given. In a steady-flow equilibrium the photoinduced e^-/h^+-pair density is constant, and thus the following continuity equation holds:

$$\frac{dn_e^{ph}}{dt} = \frac{dn_h^{ph}}{dt} = 0 \Rightarrow r_{gen} - r_{rec} - \frac{1}{e}\mathrm{div}\vec{j} = 0, \tag{10.14}$$

where $\vec{j}$ is the local current density. Under the simplified assumption of a homogeneous charge carrier density in both bands throughout the entire bulk of the absorber volume V and negligible surface effects, this leads to the following expression, linking the external current I to the generation and recombination rates, according to the divergence theorem:

$$\frac{I}{e} = V(r_{gen} - r_{rec}) =: R_{gen} - R_{rec}, \tag{10.15}$$

with R_{gen} and R_{rec} being the total generation and recombination rates of e^-/h^+-pairs in the entire absorber volume, respectively. This equation can then be normalized to the absorber area A_A, yielding

$$\frac{I}{eA_A} = \frac{R_{gen}}{A_A} - \frac{R_{rec}}{A_A}. \quad (10.16)$$

The generation rate is constant and depends only on the intensity of the solar irradiation and the geometry and material of the absorber, while the recombination rate depends on the properties of the absorber material as well as the densities of electrons and holes. For an evaluation of the operation conditions of a given solar cell under a given illumination, it is thus important to understand the basic recombination mechanisms.

While for a principle consideration most recombination mechanisms for e^-/h^+-pairs can be neglected, this is not possible for radiative recombination, i.e. the inverse process to the photoinduced generation of electrons and holes. The reason is that this process is thermodynamically linked to the light absorption process itself, similar to the heat radiation from the absorber in solarthermal devices discussed in Chapter 9. Since typical semiconductors are selective absorbers with a cut-off energy, i.e. bandgap, much larger than $3kT_{am} \approx 75$ meV, thermal radiation from them could be expected to be negligible. This would indeed be true if there were no photogenerated charge carriers present. In this case all e^-/h^+-pairs would have to be thermally generated across the bandgap, leading to a very low charge carrier density. This charge carrier density is typically denoted n_i and is called **intrinsic charge carrier density**. It is determined by the material properties of the semiconductor in the following way (for a derivation see Section B.2.1 of the appendix):

$$n_i = \sqrt{N_C N_V} e^{-E_g/2kT} \quad (10.17)$$

and is of the order of 10^{10} cm^{-3} in silicon, i.e. it is very small indeed. However, the charge-carrier density of the illuminated semiconductor is much larger, corresponding to a state far from equilibrium, and consequently there is a greatly enhanced driving force towards equilibrium via charge carrier recombination. Radiative recombination can formally be written as a reaction between an electron in the conduction band and a hole in the valence band yielding a photon:

$$e^- + h^+ \rightarrow \gamma. \quad (10.18)$$

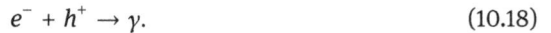

The reaction rate is then proportional to the product of the activities of the two reactants, and thus proportional to the product of their densities $n_e n_h$. Hence, it is strongly enhanced in the illuminated absorber, compared to the absorber in the dark. The enhancement factor simply reflects the change in the activities of the reactants in both cases, i.e. it is given by

$$\frac{n_e n_h}{n_i^2} = \frac{n_e n_h}{N_C N_V} \cdot e^{E_g/kT} = \exp\left(\ln\left(\frac{n_e n_h}{N_c N_v}\right) + \frac{E_g}{kT}\right)$$
$$= \exp\left(\frac{E_g}{kT} - \ln\left(\frac{N_c N_v}{n_e n_h}\right)\right) = e^{(\tilde{\mu}_{CB} - \tilde{\mu}_{VB})/kT} =: e^{\Delta\tilde{\mu}/kT}. \tag{10.19}$$

The enhancement factor, and thus the radiative recombination rate, are thus exponentially dependent on the difference in the quasi-electrochemical potentials $\Delta\tilde{\mu}$. For a given bandgap E_g the total amount of photons per unit time and absorber area stemming from radiative recombination R_{rec}/A_A is then given by the product of the area specific heat radiation at $T = T_{am}$ for a selective absorber which only emits photons with energies above E_g and the enhancement factor given in equation (10.19):

$$R_{rec}/A_A = e^{\Delta\tilde{\mu}/kT_{am}} \cdot \frac{2\pi}{h^3c^2} \int_{E_g}^{\infty} \frac{E^2}{e^{E/kT_{am}} - 1} dE$$
$$\approx e^{\Delta\tilde{\mu}/kT_{am}} \cdot \frac{2\pi}{h^3c^2} \int_{E_g}^{\infty} E^2 e^{-E/kT_{am}} dE$$
$$= e^{\Delta\tilde{\mu}/kT_{am}} \cdot \frac{2\pi}{h^3c^2} \left(E_g^2 kT_{am} + 2E_g(kT_{am})^2 + 2(kT_{am})^3\right) e^{-E_g/kT_{am}} \tag{10.20}$$
$$= \frac{2\pi}{h^3c^2} \left(E_g^2 kT_{am} + 2E_g(kT_{am})^2 + 2(kT_{am})^3\right) \cdot e^{-(E_g - \Delta\tilde{\mu})/kT_{am}}.$$

With equation (10.20), the efficiency of the solar photon collection with respect to radiative recombination η_{rad} can be given as a function of $\Delta\tilde{\mu}$ and thus of the photoinduced electron and hole densities imposed by the solar radiation. It is the ratio of the external particle current and the total generation rate of e^-/h^+-pairs in the entire absorber volume per unit time:

$$\eta_{rad} = \frac{I}{eR_{gen}} = \frac{R_{gen} - R_{rec}(\Delta\tilde{\mu})}{R_{gen}} = \frac{R_{gen}/A_A - R_{rec}/A_A(\Delta\tilde{\mu})}{R_{gen}/A_A}. \tag{10.21}$$

The efficiency η_{rad} is shown for four different bandgaps in Figure 10.5 as a function of the ratio between $\Delta\tilde{\mu}$ and E_g.

All the curves in Figure 10.5 show that the distances of the quasi-electrochemical potentials to both band edges is always several kT_{am}, and that the approximation $|E - \tilde{\mu}| > 3kT$ is thus indeed guaranteed to be valid through the inevitable radiative recombination. While η_{rad} decreases exponentially with $\Delta\tilde{\mu}$ and thus the volume densities of electrons and holes, the exergy per e^-/h^+-pairs increases linearly with $\Delta\tilde{\mu}$, as evident from equation (10.11). The interplay of both leads to an optimal choice for the stationary volume density, and thus for $\Delta\tilde{\mu}$, with respect to the exergy generated per unit absorber area and time, as shown in Figure 10.6.

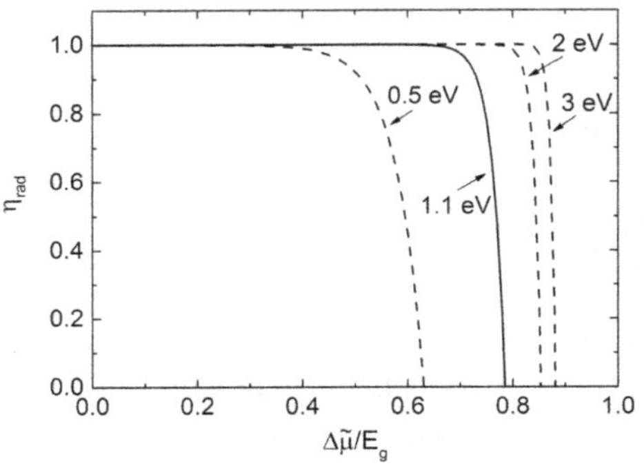

Fig. 10.5: Solar photon collection efficiency as a function of $\Delta\tilde{\mu}/E_g$ for E_g = 0.5 eV, E_g = 1.1 eV, E_g = 2 eV, and E_g = 3 eV. The *solid curve* for E_g = 1.1 eV corresponds to silicon the most important material for photovoltaic devices.

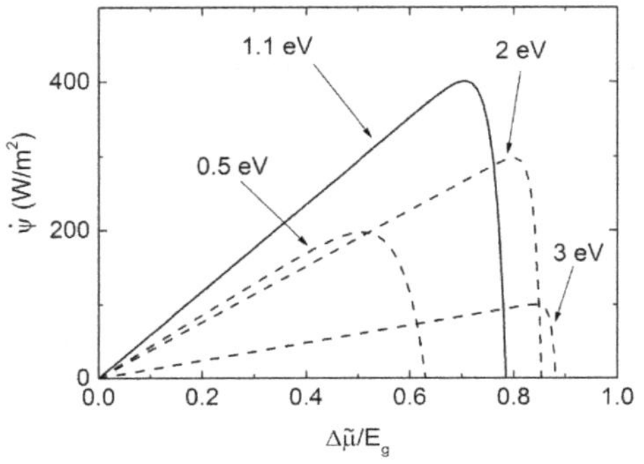

Fig. 10.6: Exergy generated per unit area and time $\dot{\psi}$ as a function of $\Delta\tilde{\mu}/E_g$ for E_g = 0.5 eV, E_g = 1.1 eV, E_g = 2 eV, and E_g = 3 eV. The *solid curve* for E_g = 1.1 eV corresponds to silicon, the most important material for photovoltaic devices.

For the optimal choice of $\Delta\tilde{\mu}$ the following values are thus found for the bandgaps given: $\Delta\tilde{\mu}_{opt}(E_g = 0.5\text{ eV}) = 0.26$ eV, $\Delta\tilde{\mu}_{opt}(E_g = 1.1\text{ eV}) = 0.78$ eV, $\Delta\tilde{\mu}_{opt}(E_g = 2\text{ eV}) = 1.6$ eV, and $\Delta\tilde{\mu}_{opt}(E_g = 2\text{ eV}) = 2.5$ eV.

10.1.3 Maximal efficiency of a solar cell

The maximal first- and second-law efficiencies of a solar cell are given by the maximal amount of electrical energy that can be taken out of the cell with incident solar irradiation. Neglecting all further loss mechanisms, this maximal electrical energy is identical to the chemical exergy produced in the bulk of the absorber. The maximal efficiency for the conversion of solar energy into chemical exergy of any semiconductor solar cell under solar black body radiation given in equation (8.5) can be decomposed into two contributions. First, the transmission of photons with an energy below the bandgap of the semiconductor leads to an efficiency contribution $\eta_{1,2,\mathrm{abs}}$, which decreases with increasing bandgap. This contribution is the ratio between the energy and exergy of the absorbed photons only and the total energy and exergy of the solar irradiation. Second, the relaxation of the photoinduced charge carriers to their energetic distribution in the steady-flow equilibrium leads to the efficiency contribution $\eta_{1,2,\mathrm{relax}}$ increasing with increasing bandgap. This is the ratio of the total extractable exergy of the relaxed charge carrier ensembles and the total absorbed energy and exergy. For the efficiency of the relaxation process, the density of minority charge carriers and thus the position of the quasi-electrochemical potentials at the end of the relaxation process is important. Thus, the emission of photons due to radiative recombination has to be taken into account when calculating this contribution.

Making use of the spectral distribution of the entropy and hence of the exergy, as given in equation (8.14), the first- and second-law efficiencies of the absorption process $\eta_{1,2,\mathrm{abs}}$ can be calculated in the following way:

$$\begin{aligned} \eta_{1,\mathrm{abs}}(E_g) &= \frac{\int_{E_g}^{\infty} E^3/(e^{E/kT_s}-1)dE}{\sigma T_{\mathrm{sun}}^4} \\ \eta_{2,\mathrm{abs}}(E_g) &= \frac{(1-T_{\mathrm{am}}/T_{\mathrm{sun}})\cdot\int_{E_g}^{\infty} E^3/(e^{E/kT_s}-1)dE}{(1-4T_{\mathrm{am}}/3T_{\mathrm{sun}})\sigma T_{\mathrm{sun}}^4} \\ &\quad - \frac{kT_{\mathrm{am}}\int_{E_g}^{\infty} E^2 \ln\left(1-e^{-E/kT_{\mathrm{sun}}}\right)dE}{(1-4T_{\mathrm{am}}/3T_{\mathrm{sun}})\sigma T_{\mathrm{sun}}^4}, \end{aligned} \tag{10.22}$$

where the numerator is the amount of energy or exergy transported by photons with an energy higher than the bandgap, while the denominator is the total energy and exergy input of the solar irradiation, respectively.

The calculation of the maximal first- and second-law efficiencies of the relaxation process is a little more complicated, and it is useful to first introduce the average energy $\bar{e}_{\mathrm{ph}}$ and exergy $\overline{\psi}_{\mathrm{ph}}$ per absorbed photon, i.e. the ratio of the absorbed energy and exergy and the number of absorbed photons both per unit time:

$$\begin{aligned} \bar{e}_{\mathrm{ph}}(E_g) &= \frac{\int_{E_g}^{\infty}(dI/dE)dE}{R_{\mathrm{gen}}/A_{\mathrm{A}}} = \frac{\int_{E_g}^{\infty} E^3/(e^{E/kT_{\mathrm{sun}}}-1)dE}{\int_{E_g}^{\infty} E^2/(e^{E/kT_{\mathrm{sun}}}-1)dE} \\ \overline{\psi}_{\mathrm{ph}}(E_g) &= \frac{\int_{E_g}^{\infty}(d\dot{\Psi}/dE)dE}{R_{\mathrm{gen}}/A_{\mathrm{A}}} \\ &= \left(1-\frac{T_{\mathrm{am}}}{T_{\mathrm{sun}}}\right)\bar{e}_{\mathrm{ph}}(E_g) - kT_{\mathrm{am}}\frac{\int_{E_g}^{\infty} E^2\ln\left(1-e^{-E/kT_{\mathrm{sun}}}\right)dE}{\int_{E_g}^{\infty} E^2/(e^{E/kT_{\mathrm{sun}}}-1)dE}. \end{aligned} \tag{10.23}$$

Here again, equation (8.14) was used for the last identity.

For the calculation of the maximal first- and second-law efficiencies of the relaxation process, $\eta_{1,\mathrm{relax}}$ and $\eta_{2,\mathrm{relax}}$, the dependence of the chemical exergy on the charge carrier density has to be considered, and the optimal value for $\Delta\tilde{\mu}_{\mathrm{opt}}$ with respect to radiative recombination has to be chosen. The total first- and second-law efficiencies of the relaxation process are then given by

$$\begin{aligned}\eta_{1,\text{relax}} &= \frac{\Delta\tilde{\mu}_{\text{opt}}}{\overline{e}_{\text{ph}}(E_{\text{g}})}\cdot\eta_{\text{rad}}(\tilde{\mu}_{\text{opt}})\\ \eta_{2,\text{relax}} &= \frac{\Delta\tilde{\mu}_{\text{opt}}}{\overline{\psi}_{\text{ph}}(E_{\text{g}})}\cdot\eta_{\text{rad}}(\tilde{\mu}_{\text{opt}}).\end{aligned} \tag{10.24}$$

The corresponding curve for the maximal possible second-law efficiency as a function of the bandgap of the absorber material is shown in Figure 10.7.

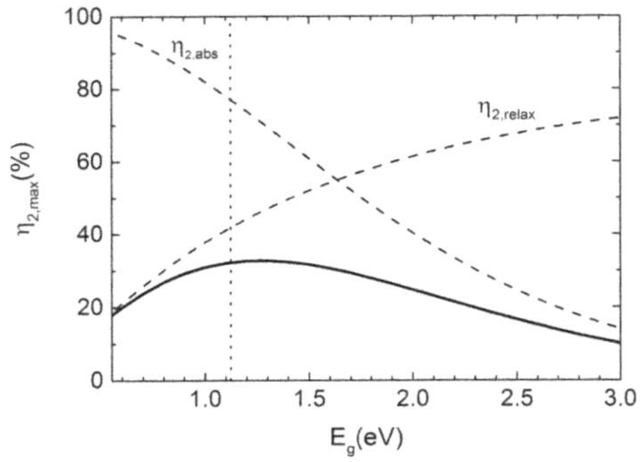

Fig. 10.7: Maximal second-law efficiency (solid) for the conversion of solar black body irradiation into chemical energy in a single junction solar cell together with the contribution $\eta_{2,\text{abs}}$ and $\eta_{2,\text{relax}}$ (*dashed curves*) as a function of the bandgap. The bandgap of silicon is indicated by the *dotted curve.*

The optimal bandgap with respect to the maximal second-law efficiency lies at ca. 1.3 eV, where the latter takes a value of $\eta_{2,\max} = 33\,\%$. Typically, the fact that the solar radiation itself is not pure exergy is neglected, and instead of the maximal second-law efficiency shown in Figure 10.7, the maximal first-law efficiency is shown. This value can be found by multiplying the curve in Figure 10.7 by 0.93, leading to a maximal possible efficiency for the conversion of solar energy into electrical exergy with a solar cell of $\eta_{\max} = 30\,\%$. This limit is often referred to as the **detailed balance** or **Shockley-Queisser limit.**

When the atmosphere of the earth is taken into account and the AM1.5 spectrum is used instead of the solar black body spectrum, the bandgap dependencies of the maximal possible efficiencies for the conversion from solar to chemical energy and exergy change. It is shown for the second-law efficiency in Figure 10.8.

In this case there is a plateau of the maximal efficiency extending, from a bandgap of 1.1 eV to 1.4 eV, with a value of $\eta_{2,\max} \approx 36\,\%$. Again, the more common term for the efficiency is the first-law efficiency, which shows an identical plateau and in this case takes a value of $\eta_{\max} \approx 34\,\%$.

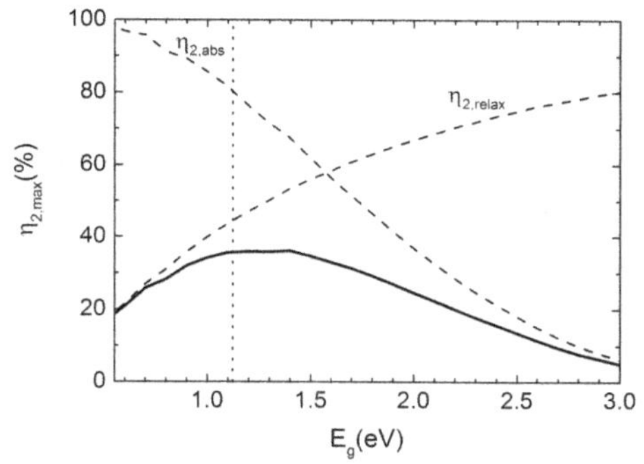

Fig. 10.8: Maximal second-law efficiency (*solid curve*) for the conversion of solar AM1.5 irradiation into chemical energy in a single junction solar cell together with the contribution $\eta_{2,abs}$ and $\eta_{2,relax}$ (*dashed curves*) as a function of the bandgap. The bandgap of silicon is indicated by the *dotted curve*.

10.2 Selective contacts

As laid out in Section 5.2, any voltage source has to be built with an intrinsic asymmetry, which leads to different electrochemical potentials in both leads of the voltage source. In a photovoltaic device, this is achieved by designing the electrical contacts of the absorber so that one contact permits only the passage of holes and the other contact only the passage of electrons. To achieve this, the electron-selective contact is realized by a material which has an electrochemical potential close to the expected value of $\tilde{\mu}_{CB}$. Conversely, the contact material of the hole-selective contact has to have an electrochemical potential close to the expected value of $\tilde{\mu}_{VB}$. In c-Si solar cells the electron selective contact is a thin ($\approx$ 10 μm) strongly n-doped region, while the hole-selective contact is strongly p-doped. Due to the n- and p-doping in the respective contacts, the electrochemical potentials are pinned there, and both quasi-electrochemical potentials from the absorber have to align in the contact region. The physical effect responsible for this alignment of the quasi-electrochemical potentials is surface recombination, i.e. an increased rate of e^-/h^+-pair recombination due to the presence of the surface. As charge carriers are constantly removed close to the surface, gradients occur in both quasi-electrochemical potentials, leading to an effective recombination current $j_{\text{surfrec,e,h}}$ at the surface. This current is identical for both types of charge carriers, because at each surface an identical amount of charge carriers from both bands is removed by surface recombination. However, as the density of the majority charge carriers is much higher than the density of the minority charge carriers, e.g. $n_\text{h}^\text{pSi} \gg n_\text{e}^\text{pSi}$ for the hole selective contact while the mobilities b_h^pSi and b_e^pSi of both charge carrier types are uniform throughout the entire solar cell, the gradient in the electrochemical potential is much steeper for the minority charge carriers. and thus for the hole selective contact, the recombination currents of electrons and holes are given by

$$
\begin{aligned}
0 &= j_{\text{surfrec,h}} + j_{\text{surfrec,e}} = -n_\text{h}^\text{pSi}(\vec{x})b_\text{h}^\text{pSi}\nabla\tilde{\mu}_\text{VB} + n_\text{e}^\text{pSi}(\vec{x})b_\text{e}^\text{pSi}\nabla\tilde{\mu}_\text{CB} \\
&|n_\text{h}^\text{pSi}(\vec{x})b_\text{h}^\text{pSi}| \gg |n_\text{e}^\text{pSi}(\vec{x})b_\text{e}^\text{pSi}| \Rightarrow |\nabla\tilde{\mu}_\text{CB}| > |\nabla\tilde{\mu}_\text{VB}|.
\end{aligned}
\tag{10.25}
$$

As a consequence, the quasi-electrochemical potentials align close to the electrochemical potential of the majority charge carriers in the doped selective-contact regions. The inequalities in equation (10.25) are inverted for n-type doped silicon where the quasi-electrochemical potentials align close to the conduction band edge. Thus, to achieve selectivity of the contacts, one contact is p-type doped and the other one n-type doped, and a pn-junction is introduced between both contacts. The charge carriers interact with the absorber at different energy levels on the two sides leading to an external voltage U_{oc}, as depicted in Figure 10.9.

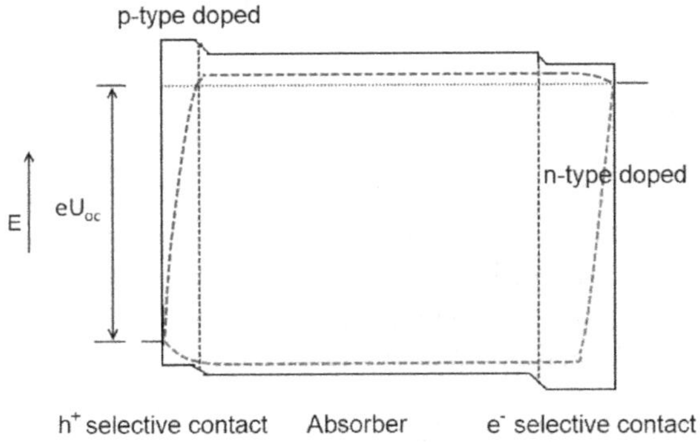

Fig. 10.9: Electron- (*right*) and hole- (*left*) selective contacts achieved by a pn-junction.

Note that the only function of the overall pn-junction is to introduce two contacts which are selective for each of the two types of charge carriers. It is thus not helpful to introduce a solar cell as an illuminated pn-junction. Again, to emphasize this, the principle behind all photovoltaic cells is the combination of an absorber with selective contacts. Whether this selectivity is achieved by a change in the doping or by other means is inessential. An illuminated pn-junction thus functions as a photovoltaic cell, but not each photovoltaic cell is an illuminated pn-junction.

The maximal external open circuit voltage U_{oc}^{max} can be calculated for the case of a negligible gradient in the electrochemical potential for the majority carriers in both selective contacts:

$$\tilde{\mu}_{CB} - \tilde{\mu}_{VB} = -eU_{max}. \tag{10.26}$$

This means that the externally usable electrical exergy per charge carrier is identical to the chemical exergy per charge carrier, as can be easily read out from equation (10.13). The selective contact is thus not linked to principally unavoidable losses.

10.3 Conversion of chemical into electrical energy

Electrical power is supplied when a current flows across an electrochemical potential difference, and is thus described by the current voltage characteristic. For a solar cell based on a pn-junction between the two selective contacts, such as, e.g. a c-Si solar cell shown in Figure 10.9, the current voltage characteristic is dominated by the current voltage characteristics of the pn-junction itself. In analogy to the double layer forming at an electrode immersed into an electrolyte, a space charge region forms at a pn-junction where the n-type doped side is positively and the p-type doped side negatively charged in the direct vicinity of the junction. The width of the space charge region is in the order of $\approx 1\ \mu m$. In the event of an external bias the majority charge carriers of both sides are either driven into the pn-junction (forward bias) or away from it (reverse bias), as shown in Figure 10.10.

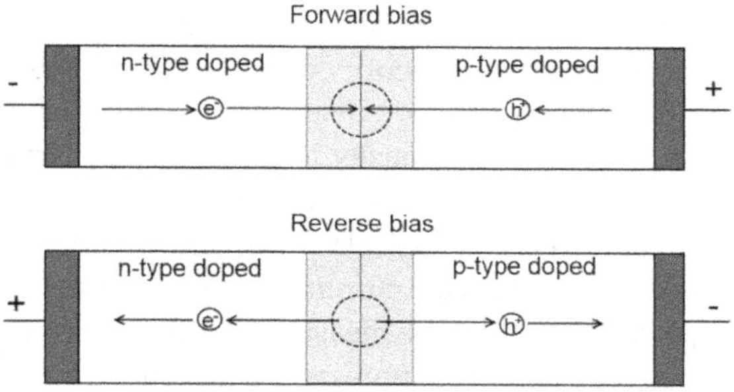

Fig. 10.10: pn-junction with forward bias (*top*) and reverse bias (*bottom*). Under forward bias the majority charge carriers of both sides recombine in the region extending one minority charge carrier diffusion length into the respective sides (*gray area*), whereas under reverse bias the charge carriers have to form thermally within this region.

The current through the bulk of the regions with different doping is carried by the majority charge carriers, as their conductivity is much higher than the one of the minority charge carriers (see equation (10.25)). At the pn-junction the charge-carrier type has to change, which is analogous to the conversion of electron conduction at an electrode surface into ion conduction in the electrolyte by means of a chemical reaction at the interface. In this case the chemical reaction is the recombination or generation reaction:

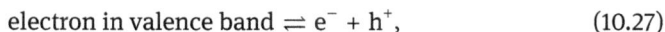

$$\text{electron in valence band} \rightleftharpoons e^- + h^+, \tag{10.27}$$

and takes place in a region extending one diffusion length of about $L_D = 10–100\ \mu m$ of the respective minority carriers into the respective regions. This spatial restriction is due to the fact that the minority carriers have a limited lifetime of about

$\tau = 100\,\mu s - 1$ ms. The diffusion length is the typical mean path length a charge carrier propagates by diffusion during its lifetime.

10.3.1 Current voltage characteristic

Under forward bias the charge carriers originate from the respective regions, where they are majority charge carriers. The transition from electron conduction in the n-type doped region to hole conduction in the p-type doped region is achieved by recombination in the reaction region. Recombination does not mean that energy is lost, as one charge carrier travels through the entire length of the pn-junction, and only its type changes by a recombination process similar to an electrochemical reaction at an electrode.

In contrast, under reverse bias, the majority charge carriers carrying the current in their respective sides originate from the region with opposite doping, where they have to be spontaneously formed by thermal excitation as minority carriers. This is a very rare process, due to the position of the electrochemical potential close to the opposite band edge. In the dark, the generated charge carrier density is the intrinsic charge carrier density n_i and thus several orders of magnitude smaller than the typical charge carrier densities generated by the doping process. An external voltage does not change this charge carrier density. Hence, under reverse bias the current is limited by the low density of thermally generated charge carriers. It can be measured for large reverse bias voltages, as the forward direction is then entirely suppressed. Consequently, it is called the dark saturation current j_ds:

$$j_\mathrm{ds} = en_\mathrm{i}^2\left(\frac{L_\mathrm{p}}{\tau_e} + \frac{L_\mathrm{n}}{\tau_h}\right) = eN_\mathrm{C}N_\mathrm{V}e^{-E_\mathrm{g}/kT}\left(\frac{L_\mathrm{p}}{\tau_e} + \frac{L_\mathrm{n}}{\tau_h}\right). \tag{10.28}$$

Here $L_\mathrm{p,n}$ are the minority charge carrier diffusion lengths in the p- and n-doped region, respectively, and $\tau_\mathrm{e,h}$ the corresponding minority charge carrier lifetimes. The kinetics of the chemical reaction in equation (10.27) determines the current voltage characteristic of the pn-junction, yielding an equation equivalent to the Butler–Volmer equation (7.35). In the direction '$\rightarrow$' of the reaction equation (10.27) the dark saturation current j_ds flows irrespective of the applied external voltage U. This is then also true in equilibrium, where identical numbers of each type of charge carrier diffuse in both directions within the "reaction zone", leading to a zero total current. This means that j_ds plays the role of the exchange current density j_0 in the Butler–Volmer equation (7.35). Furthermore, the energy barrier for the spontaneous thermal generation of electrons and holes is only determined by the bandgap and not potential dependent. According to equations (7.32) and (7.33) this leads to $\alpha = 1$, i.e. a maximally lopsided voltage response. The current voltage characteristic in the dark is then given by

$$j = j_\mathrm{ds}\left(e^{eU/kT} - 1\right). \tag{10.29}$$

When the device is illuminated, the situation changes significantly, as minority charge carriers are optically generated in the diffusion layer. While the current in the '→' direction of the reaction equation (10.27) is still independent of the voltage, its value changes significantly by the additional e^-/h^+-pair generation mechanism within the diffusion zone of the pn-junction. Under short-circuit conditions, the current corresponding to the optical generation rate of electrons and holes j_{sc} now flows across the pn-junction. It can be calculated in a similar way to j_{ds}, as it is also a diffusion current which just depends on a different charge carrier density, namely the density of photogenerated e^-/h^+-pairs $n_e^{ph} = n_h^{ph}$:

$$j_{sc} = e n_e^{ph} n_h^{ph} \left(\frac{L_p}{\tau_e} + \frac{L_n}{\tau_h} \right). \tag{10.30}$$

At the same time, at $j = 0$ an open-circuit voltage $U_{oc} \neq 0$ V adjusts between the contacts of the device. The total current voltage characteristic under illumination is thus given by

$$j = (j_{sc} + j_{ds}) \left(e^{e(U - U_{oc})/kT} - 1 \right). \tag{10.31}$$

It is possible to calculate the open-circuit voltage using an analog of the Nernst equation equation (7.16), under the assumption that the chemical activities of the electrons and holes $a_{e,h}$ are both linearly dependent on their volume densities. This assumption is plausible, because the electron and hole ensembles both behave similarly to ideal gases and can thus be seen as ensembles of noninteracting particles. The ratio of the chemical activities of the reactants on the right-hand side of equation (10.27) between the dark and the illuminated state is given by

$$\frac{a_{e,ill} a_{h,ill}}{a_{e,dark} a_{h,dark}} = \frac{n_e n_h}{n_i^2} = \frac{n_e^{ph} n_h^{ph} + n_i^2}{n_i^2} = \frac{j_{sc} + j_{ds}}{j_{ds}}. \tag{10.32}$$

Inserting equation (10.32) into the Nernst equation, equation (7.16), with $U^0 = 0$, the open-circuit voltage becomes

$$U_{oc} = \frac{kT}{e} \ln \left(\frac{j_{sc} + j_{ds}}{j_{ds}} \right). \tag{10.33}$$

Finally, using equation (10.33), the current voltage characteristic in the illuminated case as given in equation (10.31) can be rearranged:

$$j = (j_{sc} + j_{ds}) \left(\frac{j_{ds}}{j_{sc} + j_{ds}} e^{eU/kT} - 1 \right) = j_{ds} \left(e^{eU/kT} - 1 \right) - j_{sc}. \tag{10.34}$$

This means that the current voltage characteristic under illumination can also be derived by simply offsetting the current voltage characteristic in the dark by the photogenerated short-circuit current j_{sc}. Examples for current voltage characteristics in the dark and under illumination are shown in Figure 10.11 for $j_{ds} = 1 \cdot 10^{-10}$ mA/cm^2 and $j_{sc} = 40$ mA/cm^2.

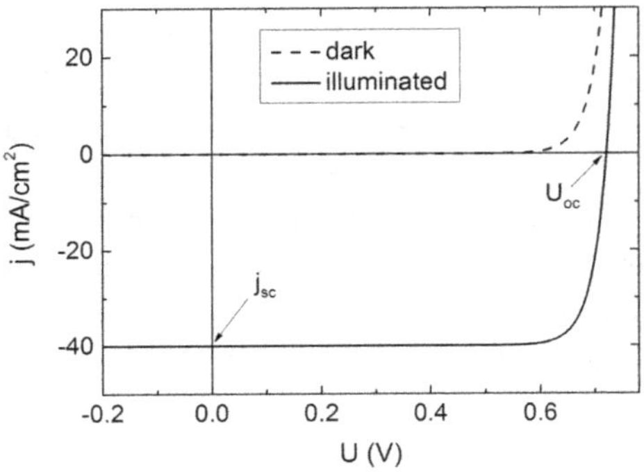

Fig. 10.11: Current voltage characteristic for a pn-junction in the dark (*dashed curve*) and under illumination (*solid curve*) with $j_{ds} = 1 \cdot 10^{-10}$ mA/cm^2 and $j_{sc} = 40$ mA/cm^2.

10.3.2 Electrical power output

Electrical power can be extracted from the solar cell in the quadrant where the voltage is positive while the current is negative, in accordance with the usual sign convention:

$$-(P/A)_{el} = U \cdot j = U \cdot \left(j_{ds}\left(e^{eU/kT} - 1\right) - j_{sc}\right). \tag{10.35}$$

The electrical power extracted is zero for both $U = 0$ and $U = U_{oc}$, which means that a maximum in power generation between these two values must exist. This maximum is called the **maximum power point** (MPP) and is characterized by the filling factor c_f, as shown in Figure 10.12:

$$c_f = \frac{U_{MPP} \cdot j_{MPP}}{U_{oc} \cdot j_{sc}}. \tag{10.36}$$

When the solar cell is perfect, and no losses occur that could have in principle been avoided, the maximum power point coincides with $U_{MPP} = \Delta\tilde{\mu}_{opt}$, as derived in Section 10.1.2. The reason for this is that every photogenerated electron either contributes to the external current or necessarily recombines radiatively, which is the exact same assumption underlying the derivation of $\Delta\tilde{\mu}_{opt}$. In this case, the diffusion lengths and lifetimes for the dark saturation current are exclusively set by radiative

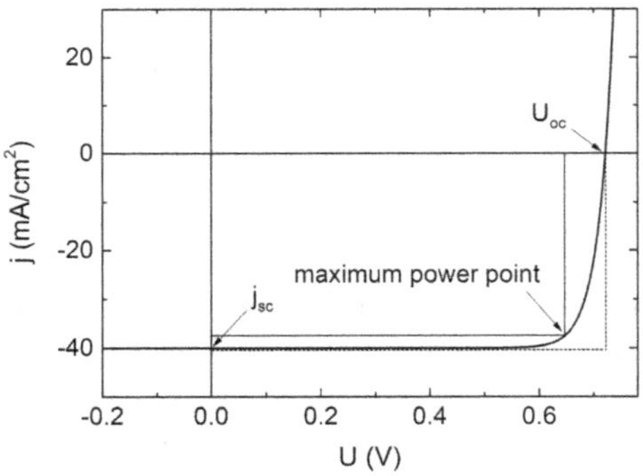

Fig. 10.12: Maximum power point for a solar cell with $j_{ds} = 1 \cdot 10^{-10}$ mA/cm^2 and $j_{sc} = 40$ mA/cm^2. According to equation (10.36) the filling factor is the area ratio between the two rectangles (*solid* and *dotted lines*) depicted.

recombination, i.e. all other recombination mechanisms present in a real photovoltaic cell are neglected. These mechanisms will be briefly introduced in the next section.

10.3.3 Losses

There are two classes of loss mechanisms in solar cells: Ohmic losses and recombination losses. As for the latter type of losses, if photogenerated charge carriers do not reach the pn-junction and are consequently recombining in the region where they are minority charge carriers, the recombination process is not associated with an external current, and the free energy is therefore lost. Recombination losses are the most important loss mechanisms in photovoltaic devices and lead to a reduction of the open-circuit voltage U_{oc} and the filling factor c_f by reducing the densities of electrons and holes and thus $\Delta\tilde{\mu}$ at a given extracted current. In typical solar cells radiative recombination (see Section 10.1.2) is not the dominant recombination mechanism. Instead, the important recombination mechanisms all proceed along a nonradiative pathway. In these processes phonons are emitted instead of photons, leading to a heating of the crystal lattice.

Ohmic losses: Due to the finite conductivity of the bulk absorber region of the solar cell, according to equation (5.12), a gradient in the electrochemical potential is necessary to drive the current through the solar cell:

$$j =: j_e = n_e b_e \cdot \nabla\tilde{\mu}_e = n_h b_h \cdot \nabla\tilde{\mu}_h =: j_h, \tag{10.37}$$

where j_e and j_h are the currents crossing the electron and hole selective contacts, respectively. This means that the external current is driven by regions where a gradient in the electrochemical potential occurs in a region with both a high mobility and a high density for the respective charge carrier type. In Figure 10.13 the gradients of the electrochemical potentials for electrons and holes responsible for an external current are shown, together with the gradients of the electrochemical potentials at the contacts which do not lead to an external current.

Note that at any given, stable external current the profile of the electrochemical potential between the contacts of the solar cell is stable with time in accordance with the notion of a steady flow equilibrium.

Surface recombination: Surface recombination currents at the selective contacts do not contribute to the external current, as can be seen in equation (10.25). The small increase of the gradient in the quasi-electrochemical potentials visible in Figure 10.13 close to the contact selective for the respective charge carrier type mark the losses due to surface recombination close to these contacts.

Impurity scattering: The important scattering mechanism at impurities or crystal imperfections and doping atoms leads to an increased recombination by slowing down electrons and holes. It is a bulk recombination process and is a function of

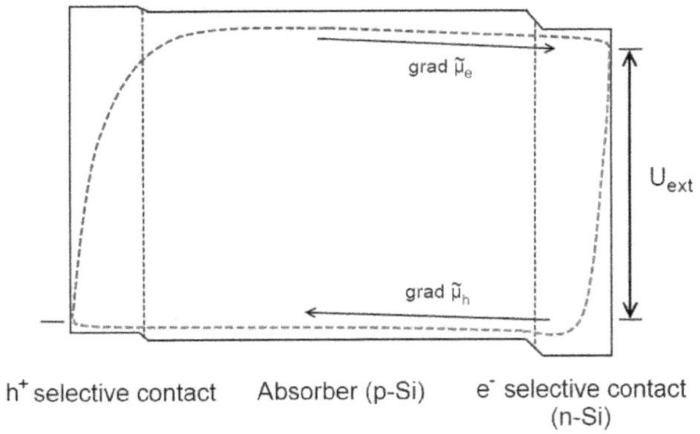

Fig. 10.13: Steady state energetic configuration of a solar cell operated at a constant current. The marked gradients of the electrochemical potentials in valence and conduction band are responsible for the ohmic losses, whereas the gradients close to the contacts are responsible for the surface recombination losses.

the crystal quality. In polycrystalline solar cells the grain boundaries offer additional recombination sites.

Auger recombination: In this bulk recombination process an e^-/h^+-pair recombines in the presence of a majority charge carrier, which first takes up the energy and then thermalizes thus converting the recombination energy to heat. The Auger recombination rate r_{rec}^{Aug} is proportional to the square of the majority charge carrier density and depends linearly on the minority charge carrier density. For holes as majority charge carriers the following is thus valid:

$$r_{rec}^{Aug} \propto n_h^2 n_e. \tag{10.38}$$

e$^-$/h$^+$/phonon scattering: Electrons and holes can be scattered by phonons, which in turn leads to an increased recombination rate. As the density of phonons is increasing with temperature, this recombination pathway leads to an increase in the overall recombination rate with temperature.

With all mechanisms discussed in the previous sections an exergy flow diagram for a solar cell as shown in Figure 10.14 is obtained.

The first two energy conversion processes represent the conversion from solar to chemical energy in the bulk of the absorber, where the first process is associated with the absorption process and the second one with the relaxation of the photogenerated charge carriers to the steady flow equilibrium. This second step involves thermalization and all recombination losses, and the filling factor. The last energy conversion step is then associated with the extraction of the charge carriers, and the only losses occurring are surface recombination and ohmic losses.

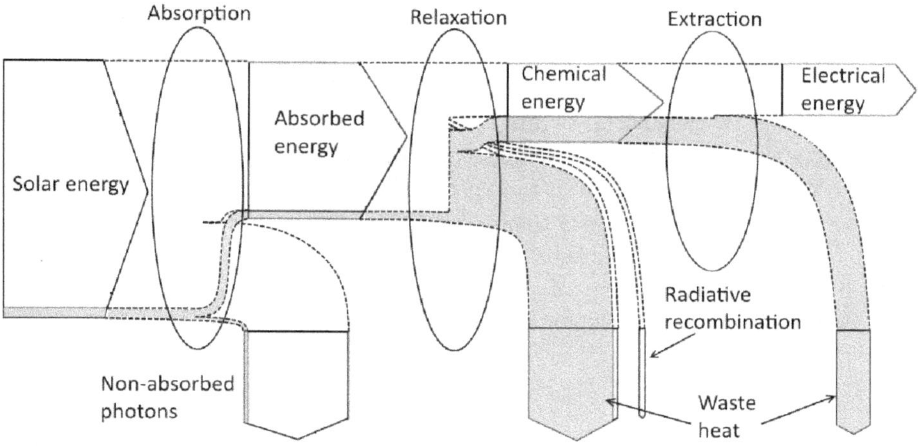

Fig. 10.14: Energy and exergy flow diagram for the energy conversion processes in a photovoltaic cell. The *ellipsoids* represent energy conversion processes and the *arrows* represent transferred energy with the *upper white fraction of the arrows* representing the transferred exergy.

10.4 Solar cell design

In this section the priciple design criteria for different types of solar cells will be briefly discussed. The main focus for each type of cell is how absorber and selective contacts are realized. As crystalline silicon solar cells are the most common type of solar cells, they will be discussed in a little more detail than the other types of solar cells.

10.4.1 c-Si solar cell

Similar to the other devices discussed so far, e.g. fuel cells in Section 7.2.4, it is instructive to consider the design requirements imposed by the various loss mechanisms. Apart from the recombination mechanisms, reflection at the front part of the solar cell leads to a decreased value of j_{sc} as less photons reach the absorber. To keep these losses at a minimum, an antireflective coating, typically Si_3N_4, is used, and the surface of the cell is roughened. The antireflective coating leads to the blueish appearance of many c-Si solar cells, as the reflectivity decreases in the red region of visible light and increases in the blue region. A second important function of the coating is to passivate surface states at the silicon surface, which leads to a decreased surface-recombination rate. Below the surface coating there is a thin region ($\approx$ 10 µm) of n-type doped silicon with a relatively high doping concentration to assure the electron selectivity of the front contact. Because silicon absorbs light poorly, as it is an indirect semiconductor, most of the solar radiation is absorbed in a lowly doped p-type region connected to the n-type region via a pn-junction. The depth of the absorber, which is several 100 µm, is chosen so that nearly all photons are absorbed before reaching the backside of the solar cell. In this light-absorbing part the recombination losses have to be minimized in order to increase the electron (minority charge carriers) lifetime and thus the probability of these charge carriers to reach the pn-junction. This is why the region is only lowly p-type doped, leading to a higher crystal quality with less scattering centers in the form of doping atoms, as well as to diminished Auger recombination. To assure efficient transport of the holes to the back contact and to increase its selectivity for holes, a thin highly p-doped layer is introduced at distances further than one electron diffusion length away from the pn-junction. A schematic of a solar cell following the mentioned design criteria is shown in Figure 10.15.

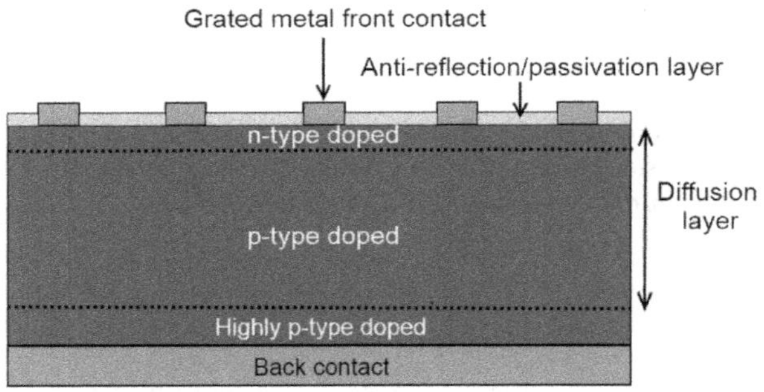

Fig. 10.15: Schematic of a c-Si solar cell.

10.4.2 Other types of solar cells

Apart from crystalline silicon solar cells, there are also a number of other types of solar cells, the most widely used types being thin film solar cells. They are made of materials which have much higher absorption coefficients than the indirect semiconductor silicon (see Section B.2.3 of the appendix). Consequently, solar radiation can be absorbed in a film with a thickness of only a few micrometers. This film can be produced by evaporation or printing techniques, which, together with the small amount of material required, makes the production process relatively inexpensive. The front contacts are typically made of a transparent conductive oxide (TCO), as, e.g. aluminum-doped ZnO, instead of a grated contact, as in crystalline silicon solar cells.

As measured by the market penetration, the most relevant absorber materials are inorganic direct semiconductors, such as copper/indium/gallium/selenide (CIGS) alloys or also amorphous silicon (a-Si). Depending on the indium/gallium ratio, the former have a bandgap tunable between 1.0–1.7 eV and the latter a bandgap around 1.5 eV, which means that both are well suited for the solar spectrum. The main drawback of the CIGS solar cells is the rarity of the materials indium, gallium, and selenium, and for the a-Si cells a relatively low efficiency.

An alternative class of absorber materials are organic semiconductors. They typically consist of blends of electron- and hole-conductive polymers, each linked to one of the electrical contacts. The main optimization task for this type of solar cells is the e^-/h^+-pair separation, as electrons and holes have very small minority charge carrier life-times and diffusion lengths. While the efficiency of organic solar cells is currently not comparable to other technologies, their inexpensive production costs and abundance of the materials required makes them a field of intense research.

Finally, there are dye-sensitized solar cells, which use an organic dye as the absorber. This dye is in contact with a high surface area n-doped TiO_2 electrode as the electron selective contact, and an electrolyte typically containing the chemical species I_2/I_3^- as the hole selective contact:

$$2I_3^- + 2h^+ \rightarrow 3I_2. \tag{10.39}$$

This electrolyte is contacted at the other side via another electrode, where the opposite direction of reaction equation (10.39) takes place and the triiodide (I_3^-) ions thus act as a charge shuttle through the electrolyte. Dye-sensitized solar cells currently show higher efficiencies than organic solar cells, but are nevertheless not widely used in commercial applications.

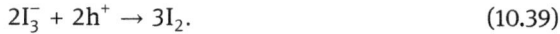

11 Exergy storage

In the previous chapters, the generation of electrical energy from different sources was discussed. The focus was a deepening of the understanding of energy conversion processes as a basis for a well-informed evaluation of different processes discussed within the scientific community and by the general public. In this context, one should further consider the requirements of a sustainable energy infrastructure. For this, it is imperative to decrease the dependence on finite resources, most importantly fossil fuels. Not only are these fuels not a durable source of exergy, their combustion also increases the concentration of the greenhouse gas CO_2 in our atmosphere, possibly severely interfering with the standard of living of large parts of humanity and the biosphere in general. From this viewpoint, on the one hand a drastic reduction of the primary energy consumption by means of more efficient converters and a less excessively energy-consuming economy are required. On the other hand, the primary energy consumption which is then still necessary should be exclusively derived from renewable sources. In general, these energy sources fluctuate in time. For this reason, exergy storage systems play an important role in any conceivable energy infrastructure revolving around renewable energies, as otherwise the production of energy cannot always be adapted to the current demand. Thus, a brief overview over different possibilities for exergy storage is given in this chapter, whereby the concepts and equations derived in the previous chapters will be applied.

The exergy storage devices discussed are either genuine storages for electrical energy, i.e. with an input and an output of electrical energy, or devices that convert electrical energy into other forms of final energy, e.g. chemical fuels for mobility applications or again for electrical energy production. In addition to efficiencies, as important characteristic numbers for the storage devices, the volume energy densities and typical timescales for the storage, from seconds to months, are discussed.

11.1 Mechanical exergy storage

The storage of electrical energy in a mechanical way is not only the first possibility that comes to mind, but also, as conversions into lower exergy content energy forms do not take take place, a path promising low overall losses. Two main routes are chosen in this context: **pumped storage hydro power plants** and **compressed air energy storage**. Both techniques are applied on large scales in the several GWh storage capacity range.

11.1.1 Pumped storage hydro power plant

Pumped hydro power is the most straight forward way of storing electrical energy on a large scale. It is the only large scale exergy storage technology already in wide use. Here, water is pumped over a height difference Δh from a lower to an upper water reservoir. As seen in Section 3.3.2 the potential energy of water is pure exergy. This means that the total first- and second-law efficiencies are identical and solely determined by the first- and second-law efficiencies of the water pumps and turbines. Realistically, the second-law efficiency of the combined storage/release process is relatively high and lies above 80 %. Furthermore, this storage method is very flexible with respect to storage time and power density, the former ranging from minutes to months and the latter typically being in the order of 100 MW. These positive aspects are only counteracted by the main drawback of this form of energy storage, the relatively low exergy volume density Ψ/V. At an assumed Δh = 100 m it is only

$$\Psi/V = 0.27\ \text{kWh/m}^3, \tag{11.1}$$

leading to the need for massive lakes as reservoirs.

11.1.2 Compressed air energy storage

Beside pumped storage hydro power plants, the only other large-scale exergy storage facilities already in use, albeit on a smaller scale, are compressed air energy storage (CAES) facilities. In such facilities air is compressed with surplus electrical energy and pumped into large subterranean caverns. This compressed air can then be used together with auxiliary gas to drive a gas turbine, as schematically shown in Figure 11.1.

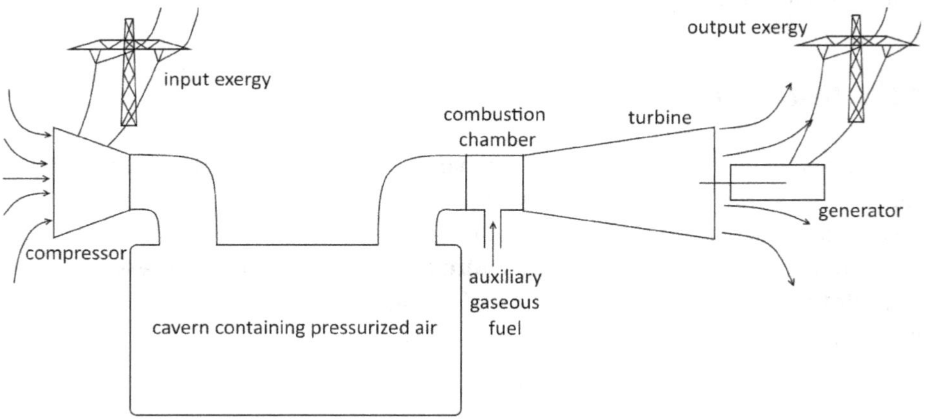

Fig. 11.1: Schematic of a compressed air energy storage facility.

A CAES facility thus has the advantage that it can be started very quickly, within seconds. The input exergy density per unit mass Ψ_{in}/m is characterized by the initial and final pressure $p_{i,f}$ of the storage volume. Hence, for an adiabatic compression, the exergy input per unit mass, Ψ_{in}/m, is given by

$$\frac{\Psi_{in}}{m} = \int_{p_i}^{p_f} v dp = \int_{p_i}^{p_f} \frac{p_i^{1/\gamma}}{\varrho_i p^{1/\gamma}} dp = \frac{\gamma}{\gamma - 1} \frac{p_i^{1/\gamma}}{\varrho_i} \left(p_f^{1-1/\gamma} - p_i^{1-1/\gamma}\right). \tag{11.2}$$

Under the assumption of an ideal gas with an adiabatic exponent of $\gamma = 7/5 = 1.4$, and at a typical final pressure of $p_f = 50$ bar, this leads to the following mass-specific input exergy:

$$\frac{\Psi_{in}}{m} = 610 \text{ J/g}. \tag{11.3}$$

The main drawback of this exergy storage technique is the problem of heat insulation. In the example just considered the compressed gas is heated to $T_f = 910$ K $= 637$ °C. Conversely, the gas in the cavern does not have a temperature much higher than $T = T_{am}$ after some storage time. This leads to an isochoric cooling of the compressed gas. Under the assumption of a constant mass-specific heat capacity $c_v = 0.72$ J/(gK), the corresponding maximal storage mass-specific loss of exergy Ψ_{loss}/m can be calculated according to equation (3.21):

$$\Psi_{loss}/m = c_v \cdot \left((T_f - T_{am}) - T_{am} \cdot \ln\left(\frac{T_f}{T_{am}}\right)\right) = 200 \text{ J/g}. \tag{11.4}$$

This means that about the third part of the exergy is lost due to the cooling of the compressed gas. The volume specific value for the exergy stored in the cavern at ambient temperature Ψ/V is then given by

$$\Psi/V = \varrho\left(\frac{\Psi_{in}}{m} - \frac{\Psi_{loss}}{m}\right) = 25 \text{ MJ/m}^3 = 6.8 \text{ kWh/m}^3. \tag{11.5}$$

This is a considerably higher value than what is achievable with a pumped storage hydro power plant. To circumvent some of the losses, CAES facilities which cool the gas during the compression process and store the released heat in a separate heat storage are being investigated, but so far are not in use. For large systems the typical discharge power is of the order of 100 MW, which can then be maintained for a few hours, depending on the storage size. In principle, a CAES can be used for all relevant timescales ranging from seconds to months.

11.2 Thermal exergy storage

There are two qualitatively different applications for thermal exergy storages. First, as already mentioned in the context of solarthermal power plants in Section 9.2.4, the thermal storage device can be aimed at driving a heat engine releasing electrical exergy and, second, its desired output can be low temperature heat or warm water for, e.g. households. In the first case, the temperature of the heat storage has to be relatively high ($T > 300\,^\circ\mathrm{C}$), while in the latter case the storage temperature can be much lower ($T < 100\,^\circ\mathrm{C}$). The heat is typically stored in a large tank filled with one medium which either increases its temperature from the heat input or changes its phase. The former is called **heat capacity storage** and the latter **latent heat storage.**

11.2.1 High temperature storage

In solarthermal power plants, typically molten-salt heat capacity storages are used. To avoid a continuous temperature spectrum of the tank the molten salts are kept in two separate containers at two distinct temperature levels T_c and T_h, as shown in Figure 9.11. The storage is charged by extracting molten salt from the low temperature tank at $T = T_c$, heating it up to the high temperature level $T = T_h$, and feeding it into the high temperature tank all isobarically. For the discharge this process is reversed. A typical high-temperature heat storage is shown schematically in Figure 11.2.

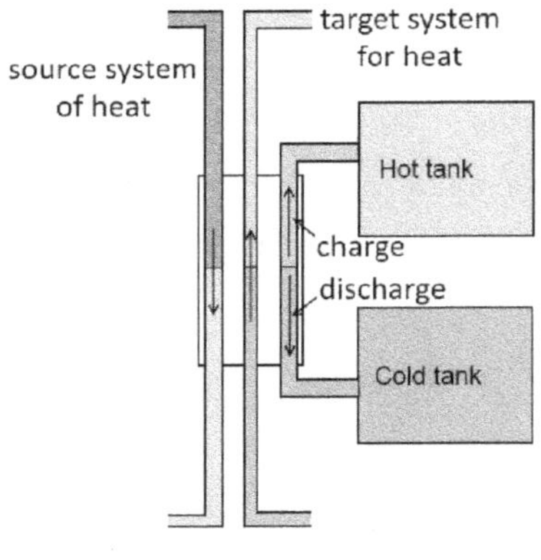

Fig. 11.2: Schematic of a high-temperature heat storage.

Under the assumption of a constant heat capacity of the molten salts, the stored exergy per unit storage volume Ψ_{st}/V is then given by equation (3.21):

$$\Psi_{st}/V = \frac{1}{2}\varrho c_p \cdot \left((T_h - T_c) - T_{am} \cdot \ln\left(\frac{T_h}{T_c}\right)\right). \tag{11.6}$$

The factor $1/2$ on the right-hand side takes into account the fact that two tanks are needed for the storage. With the temperatures $T_c = 563\ \mathrm{K} = 290\,^\circ\mathrm{C}$ and $T_h = 663\ \mathrm{K} = 390\,^\circ\mathrm{C}$, a heat capacity of $c_p = 1.5\ \mathrm{J/(gK)}$, and a volume density of the molten salt of $\varrho = 1.0\ \mathrm{g/cm^3}$ as realistic working parameters, this leads to a storage volume specific exergy of

$$\frac{\Psi}{V} = 11\ \text{kWh/m}^3. \tag{11.7}$$

The timescale for the storage is determined by the quality of the heat insulation alone. In solarthermal power plants the tanks serve the purpose of operating the power plant during the night and are consequently aimed at storage durations of ca. 1 d. The second-law efficiency of this type of exergy storage is determined by the second-law efficiency of the two heat exchange processes into and out of the heat storage. Assuming a constant temperature difference of 100 K for the heat exchanger, the heat source charging the heat storage runs between $T_{1,\text{src}} = 713$ K and $T_{2,\text{src}} = 613$ K, while the target system runs between $T_{1,\text{tar}} = 513$ K and $T_{2,\text{tar}} = 613$ K. The second-law efficiency of the heat exchange between both systems mediated by the heat storage can then, under the assumption of a constant heat capacity in both source and target system, be calculated using equation (3.25):

$$\eta_2 = \frac{1 - T_{\text{am}}/\Delta T_{\text{tar}} \cdot \ln(T_{2,\text{tar}}/T_{1,\text{tar}})}{1 - T_{\text{am}}/\Delta T_{\text{src}} \cdot \ln(T_{2,\text{src}}/T_{1,\text{src}})} = 0.85. \tag{11.8}$$

This relatively high second-law efficiency shows that salt tanks are indeed a good choice for solarthermal power plants.

11.2.2 Low-temperature storage

If the desired form of output energy is low-temperature heat, typically liquid water is used as the storage medium because of both its abundance and its high heat capacity of $c_p = 4.2\ \text{J/(gK)}$. As exergy sources, either solarthermal heat from flat panel collectors or electrical energy driving a heat pump can be used. The latter case will be used as an example for the rest of the considerations in this section. With a maximal temperature of $T_h = 363\ \text{K} = 90\ °\text{C}$, the exergy stored per unit volume is given by equation (3.21):

$$\Psi_{\text{st}}/V = \varrho c_p \cdot \left((T_h - T_{\text{am}}) - T_{\text{am}} \cdot \ln\left(\frac{T_h}{T_c}\right) \right) = 7.2\ \text{kWh/m}^3 \tag{11.9}$$

The output energy of such a low temperature heat storage is also low temperature heat q transferred in an isobaric process with an assumed realistic value of again $T_{\max} \approx 70\ °\text{C} = 343$ K for room heating and warm water generation. Its exergy content ψ/q can also be calculated using equation (3.21) and has a value of $\psi/q = 6.9\ \%$. The total amount of heat that can be rejected per unit storage volume Q/V by a low temperature energy storage water tank is thus rather high:

$$\frac{Q}{V} = \frac{\Psi_{\text{st}}}{V} \cdot \frac{q}{\psi} = 104\ \text{kWh/m}^3. \tag{11.10}$$

The drawback of low-temperature heat storage is that the exergy transfer from the storage tank to the water to be heated is only possible as long as the temperature T_{stor} of the former exceeds the desired maximal temperature T_{max} of the latter. Thus, without a further source of exergy, or a second tank, as in the case of high-temperature heat storage, only a small fraction of the exergy stored in the tank can be used. A convenient choice for an auxiliary exergy source is a second electrical or absorption heat pump which uses the low-temperature heat storage as its cold side. In this way the pumped heat is not pure anergy, increasing the overall possible first-law efficiency of the heat pump. Such a configuration is schematically shown in Figure 11.3.

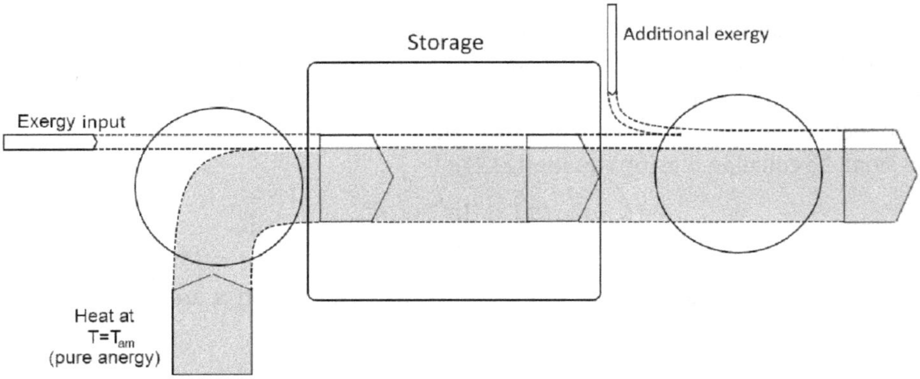

Fig. 11.3: Schematic of a low-temperature heat storage. The two heat pumps for charging *(left)* and discharging *(right)* of the storage are represented by their respective exergy flow diagrams. The temperature of the storage is below the end temperature of the heated water on the right-hand side, as evident from its lower exergy content.

The overall electrical energy saved due to the use of the storage as a cold side for the discharge heat pump, in comparison to a similar process with the ambient surroundings as the cold side, is then identical to the stored exergy given in equation (11.9). If, as is typically the case, the low-temperature heat storage is used as a seasonal exergy storage, an additional advantage arises. As the charging takes place during the summer months and the discharging during the winter months, the ambient surroundings cool down between both processes, leading to an effective upgrade of the stored energy.

11.3 (Electro-)Chemical exergy storage

Electrochemical or purely chemical energy storages are often seen as optimal energy storage devices. The reason for this is the comparatively high energy and exergy density of chemical fuels, the high possible output power, and the variable storage interval covering all relevant timescales, from seconds to months or even years. For electrochemical energy storage, a setup inverse to the galvanic cell described in Section 7.2 is used, the **electrolytic cell**. As opposed to a galvanic cell, which converts chemical energy into electrical energy, an electrolytic cell does the opposite, generating chemical energy, i.e. separated chemical fuels, making use of an external supply of electrical energy. The prototypical example is the conversion of water into hydrogen and oxygen. In such an electrolytic cell, the open-circuit electrode potentials are given by equation (7.15). Under operation conditions, i.e. when an external current is flowing, the electrode potentials at both electrodes are changing, due to overpotentials caused by the same phenomena described in detail in Section 7.2.2, which are required to drive the reaction at a nonzero speed. The effect of these overpotentials at the electrode potentials is shown in Figure 11.4.

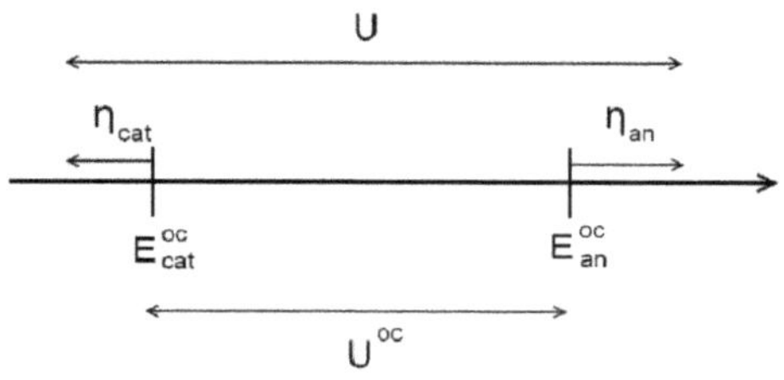

Fig. 11.4: Effect of the overpotentials in an electrolytic cell.

For the total second-law efficiency of the storage device, the succession of an electrolytic reaction, as for example water splitting, and the efficiency of the conversion of the chemical fuels into exergy have to be taken into account. The total electrochemical second-law efficiency of the storage process is thus given by the product of the second-law efficiency of the electrolytic reaction and the second-law efficiency of the inverse process, the fuel combustion reaction. It is visualized for an electrolytic cell/galvanic cell system in the full UI-characteristic in Figure 11.5.

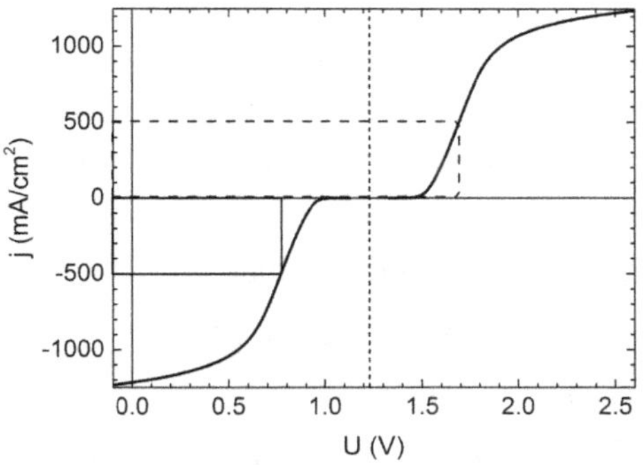

Fig. 11.5: Electrical power input (*area of dashed rectangle*) and output (*area of solid rectangle*) of an electrolytic/galvanic cell energy on-demand system.

Note that in this case, as for all storage devices, the input and output power do not necessarily match. The total second-law efficiency for the full conversion of electrical energy back to electrical energy is a function of both the input and the output power and can be read out from the ratio of the two widths of the rectangles.

11.3.1 Rechargeable batteries (Li-ion batteries)

Rechargeable batteries in general, and lithium ion batteries (Li-ion batteries) in particular, are an especially elegant way to store electrical energy in a purely electrochemical fashion. Due to the relatively high energy mass density, Li-ion batteries are well suited for portable electronic devices and also electromobility applications. The schematic of a typical Li-ion battery is shown in Figure 11.6.

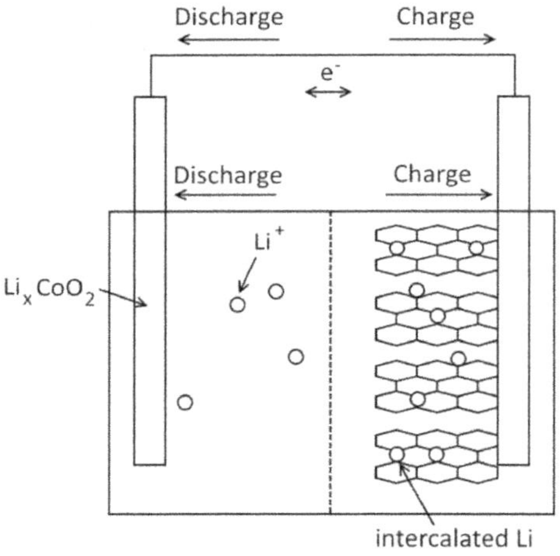

Fig. 11.6: Schematic of a conventional Li-ion battery. In the left-hand electrode $0 \leq x \leq 1$ holds. The right-hand electrode is typically made of a graphite network where the Li-ions can form bound complexes in a process called intercalation.

Li^+ ions are the mobile species within the cell. During the charging process they are reversibly dissolved from, e.g. a $LiCoO_2$ anode, travel to the other electrode, and form intercalated complexes in a graphite matrix. The two half-reactions and the overall chemical reaction thus read

$$\begin{aligned} &\mathrm{Li}_{x_1}\mathrm{CoO}_2 \rightleftharpoons \mathrm{Li}_{x_2}\mathrm{CoO}_2 + \Delta x\mathrm{Li}^+ + \Delta x e^-: && \text{left-hand electrode,}\\ &(\mathrm{C}_6)_n + \Delta x\mathrm{Li}^+ + \Delta x e^- \rightleftharpoons \mathrm{Li}_{\Delta x}(\mathrm{C}_6)_n: && \text{right-hand electrode,}\\ &\mathrm{Li}_{x_1}\mathrm{CoO}_2 + (\mathrm{C}_6)_n \rightleftharpoons \mathrm{Li}_{x_2}\mathrm{CoO}_2 + \mathrm{Li}_{\Delta x}(\mathrm{C}_6)_n: && \text{full reaction,} \end{aligned} \tag{11.11}$$

where $\Delta x = x_1 - x_2$ is the change in the Li content in the left-hand electrode, and $(C_6)_n$ represents the carbon matrix. The electrochemical potential of the Li^+ ions in both electrodes is stable over a wide range of x, and so is then also the output voltage. Typical cell voltages for the Li-ion battery lie in the region of $U^{oc} \approx 3.5$ V. The main advantage of Li-ion compared to other types of batteries is their relatively high stored exergy mass density, which lies in the order of $\Phi/m > 0.1$ kWh/kg. The main drawbacks of this type of battery are that they can relatively easily overheat and burst into flame, and that lithium is a rare material.

11.3.2 Water splitting

Storing energy in the form of hydrogen is a pathway that has been discussed for decades in the context of both electrical energy storage and mobility applications. It involves two major steps: first, the production of hydrogen from water and, second, the storage of hydrogen in a tank. Oxygen, which is also produced, is typically not stored. The reason for this is the small exergy loss caused by the dilution of the oxygen from its pure form at $p_{ox} = 1$ bar to its atmospheric concentration of $p_{ox} = 0.2$ bar. This loss is ca. 4 kJ/mol, reducing the total exergy of the reactant fuels of $\Delta g_r^0 = -475$ kJ/mol by less then 1 %. Hydrogen is produced in the inverse process of the PEM fuel cell process discussed in Section 7.2.4, the electrolysis of water or **water splitting**. The two half reactions are

$$\begin{aligned} &\text{Anode:} \quad 2H_2O \rightarrow 4H^+ + 4e^- + O_2 \; ; \; E_{el} = 0.82 \text{ V vs. SHE,} \\ &\text{Cathode:} \; 4H^+ + 4e^- \rightarrow 2H_2 \; ; \; E_{el} = -0.41 \text{ V vs. SHE,} \end{aligned} \tag{11.12}$$

where the electrode potentials are given for an electrolyte with a pH value of 7. The minimum potential difference between the two electrodes necessary to drive this reaction is the corresponding open-circuit cell voltage $U^{oc} = 1.23$ V, as given in equation (7.19). The second-law efficiency of the water-splitting process itself when only the hydrogen is captured is then given by

$$\eta_{2,\text{WS}} = \frac{\Delta g_r^0 + RT \cdot \ln\left(a_{O_2}\right)}{4FU} = \frac{\Delta g_r^0 + RT \cdot \ln\left(a_{O_2}\right)}{\Delta g_r^0 + 4F(\eta_{an} - \eta_{cat})}, \tag{11.13}$$

where $a_{O_2} = 0.2$ is the activity of the atmospheric oxygen. The main exergy losses occur due to the loss mechanisms giving rise to the overpotentials at both electrodes. For the storage of hydrogen one of two routes is typically taken. Either the gas is pressurized and stored in appropriate containers at ca. 800 bar, or the hydrogen is cooled down at atmospheric pressure below its boiling point of 21 K. In both cases the volume density of the hydrogen is similar and, in the pressurized case, the total exergy volume density Φ/V of the stored hydrogen for $p = 800$ bar is given by

$$\begin{aligned} \Phi/V &= \frac{p}{RT}\Phi_m = \frac{p}{RT_{am}}\left(\Delta g_r^0 + RT_{am} \cdot \ln\left(a_{O_2}^{1/2}\right)\right) = \frac{p}{RT_{am}}(239 \text{ kJ/mol}) \\ &= 2.1 \text{ MWh/m}^3. \end{aligned} \tag{11.14}$$

Pressurizing the hydrogen in an adiabatic compression process requires an extra exergy input Ψ_{comp}/V per unit storage volume of

$$\Psi_{comp}/V = \frac{p_{max}}{RT} \int_{p_0}^{p_{max}} v dp = \frac{p_{max}}{RT} \int_{p_0}^{p_{max}} \frac{p_0^{1/\gamma} v_0}{p^{1/\gamma}} dp$$
$$= \frac{p_{max}}{RT} \frac{\gamma}{\gamma - 1} p_0^{1/\gamma} v_0 \left(p_{max}^{1-1/\gamma} - p_0^{1-1/\gamma}\right) = 400 \text{ kWh/m}^3, \quad (11.15)$$

where v_0 is the mole-specific volume of the hydrogen in the standard state. This means that this process requires about 20 % of the exergy stored in the fuel. The exergy loss is even higher for a H_2 liquefaction process, where the total exergy required is typically ca. 25 % of the exergy stored.

11.3.3 Power to gas

Industrialized countries usually have two important power grids installed, the electrical power grid and the natural gas grid. As the latter is easily accessible, the conversion of surplus renewable energy into methane gas is an option which greatly facilitates the storage of the chemical exergy carrier. This conversion is a two step process, where the first step is the electrolysis of water to H_2 and O_2 gas, as described in the previous section. The hydrogen gas can then be further utilized to convert CO_2 into methane according to the exothermal **Sabatier process**:

$$4H_2 + CO_2 \rightarrow CH_4 + 2H_2O \;; \Delta h_r^0 = -165 \text{ kJ/mol}. \quad (11.16)$$

The maximal first- and second-law efficiencies of this conversion process can be calculated by comparing the energy and exergy input in form of $4H_2$ to the energy and exergy output in form of CH_4:

$$\eta_1 = \frac{\Delta H_{r,CH_4 \to CO_2}}{4 \cdot \Delta H_{r,H_2 \to H_2O}} = \frac{802.2}{967.2} = 83\ \%$$
$$\eta_2 = \frac{\Delta G_{r,CH_4 \to CO_2}}{4 \cdot \Delta G_{r,H_2 \to H_2O}} = \frac{800.8}{914.2} = 88\ \%. \quad (11.17)$$

Here the combustion of methane is compared to the combustion of H_2, i.e. the product water is in both cases assumed to be gaseous. In a real process, the exergy and energy conversion efficiencies are much lower than the maximal values given in equation (11.17). A major difficulty in a power to gas process is the use of CO_2 as a reactant, since it has a very low concentration in the atmosphere. The source of CO_2 for this process is typically concentrated CO_2 from the flue gas of a power plant or a bio-gas converter.

A. Basics: Thermodynamics

The entire concept of the present book centers around thermodynamics. Therefore, in this chapter a brief overview over some fundamental aspects of thermodynamics is given. Together with Section 2.1, it is aimed at providing a comprehensive basis in terms of the nomenclature and concepts used throughout this book.

A.1 Thermodynamic potentials

Thermodynamic potentials are energy functions of a thermodynamic state. They are extensive state variables characterizing the energy of a given state in a given kind of system and problem.

A closed system with m particle species can be well characterized by the **internal energy** $U(S, V, N_k, k = 1, \ldots, m)$, i.e. the sum of the energies of all particles in the system. Its total differential reads

$$dU = TdS - pdV + \sum_k \mu_k dN_k, \tag{A.1}$$

where the μ_k are the chemical potentials of the m particle species.

The **enthaply** H of a system is defined as

$$\begin{aligned} H &= U + pV \\ \Rightarrow \quad dH &= dU + pdV + Vdp = -pdV + TdS + \sum_k \mu_k dN_k + pdV + Vdp \\ &= TdS + Vdp + \sum_k \mu_k dN_k. \end{aligned} \tag{A.2}$$

As evident from the total differential, the enthalpy is a function of the variables S, p, and N, i.e. $H = H(S, p, N_k, k = 1, \ldots, m)$. The enthalpy is of special importance for open systems in a steady-flow equilibrium, as the matter constituting the system is continuously replaced and the work $pV =: mpv = mp/\varrho$ has to be provided to the system to add the mass m to the system.

A difference in the **Helmholtz free energy** F gives the maximal extractable work of a closed system undergoing an isothermal and isochoric state change. Due to the second law of thermodynamics the Helmholtz free energy is minimized in a closed system in thermal equilibrium with its surroundings. The Helmholtz free energy is the entropy free part of the internal energy U:

$$\begin{aligned} F &= U - TS \\ \Rightarrow \quad dF &= TdS - pdV + \sum_k \mu_k dN_k - TdS - SdT = -SdT - pdV + \sum_k \mu_k dN_k. \end{aligned} \tag{A.3}$$

From the total differential $F = F(T, V, N_k, k = 1, \ldots, m)$ can be read out.

The **Gibbs free energy** G plays an analogous role to the Helmholtz free energy for closed systems undergoing an isothermal and isobaric state change. A difference in the Gibbs free energy gives the maximal extractable work. It is also important for open systems in a steady-flow equilibrium. It is the entropy-free part of the enthalpy H:

$$\begin{aligned} & G = H - TS = U + pV - TS \\ \Rightarrow \quad & dG = TdS + Vdp + \sum_k \mu_k dN_k - TdS - SdT = -SdT + Vdp + \sum_k \mu_k dN_k. \end{aligned} \tag{A.4}$$

From the total differential $G = G(T, p, N_k, k = 1, \dots, m)$ is evident.

A.2 Ideal gases

An ideal gas is an ensemble of noninteracting identical particles like atoms or molecules considered as points, i.e. occupying no volume. The state of motion of the particles only changes due to interactions with the walls that define the spatial extension of the system considered. Gases are typically well described by these assumptions, and where they are valid, the properties of the system are well described by the ideal gas equation

$$pV = nRT. \tag{A.5}$$

Here n is the amount of substance measured in mole and $R = 8.314$ J/(mol K) the ideal gas constant. The equipartition theorem states that for every quadratic degree of freedom the internal energy is increased by $1/2kT$. As in an ideal gas the particles do not interact with each other; the degrees of freedom are first the translational center of mass movements and, second, those of inner motions of the molecules described by harmonic potentials, i.e. rotations and vibrations. These degrees of freedom are all of a quadratic form leading to the important relation

$$U = \frac{f}{2} NkT = \frac{f}{2} nRT, \tag{A.6}$$

where the number of quadratic degrees of freedom per molecule is denoted by f. In general the number of degrees of freedom f in a molecule is only dependent on the number of atoms N_{at} composing the molecule:

$$f = 3N_{at}. \tag{A.7}$$

There are always three translational degrees of freedom. The number of rotational degrees of freedom depends on the molecule structure and is 0 for atoms, 2 for linear molecules, and 3 for all other molecular structures. This leads to the following expressions for the number of vibrational degrees of freedom for the different types of molecules:

$$\begin{aligned} &\text{Atoms:} && f_{vib} = 0 \\ &\text{Linear molecules:} && f_{vib} = 3N_{at} - 5 \\ &\text{Other molecule structures:} && f_{vib} = 3N_{at} - 6. \end{aligned} \tag{A.8}$$

From the quantum mechanical description of the molecules it can be derived that the rotational and vibrational degrees of freedom need a certain activation energy to be excited, which means that they are only relevant above a certain temperature. This temperature is different for the distinct degrees of freedom and typically lower for rotations than for vibrations leading to a step-like increase of f with temperature.

A.2.1 Heat capacity

According to the first law of thermodynamics the internal energy of a closed system changes due to heat transfers into the system. If the volume is kept constant during the heating process no work is applied to the system, and the heat capacity at constant volume c_V can be calculated for an ideal gas as follows:

$$c_V = \left(\frac{\delta Q}{\partial T}\right)_V = \left(\frac{\partial U}{\partial T}\right)_V = \frac{f}{2}nR. \tag{A.9}$$

The heat capacity is proportional to the number of degrees of freedom active in the system. If the pressure instead of the volume is kept constant, the volume work $W = -\int pdV$ is provided to the system in addition to the heat. The heat capacity at constant pressure c_p of an ideal gas then becomes:

$$\begin{aligned} c_p &= \left(\frac{\delta Q}{\partial T}\right)_p = \left(\frac{\partial U}{\partial T}\right)_p + p\left(\frac{\partial V}{\partial T}\right)_p = \left(\frac{\partial H}{\partial T}\right)_p = \left(\frac{\partial (U+pV)}{\partial T}\right)_p \\ &= \left(\frac{\partial (U+nRT)}{\partial T}\right)_p = \left(\frac{f}{2}+1\right)nR. \end{aligned} \tag{A.10}$$

The ratio of both quantities is called the **adiabatic exponent** $\gamma := c_p/c_V$. For ideal gases it thus has a value of

$$\gamma = \frac{c_p}{c_V} = \frac{f/2+1}{f/2} = 1 + \frac{2}{f} \tag{A.11}$$

and shrinks with the number of quadratic degrees of freedom f.

A.2.2 State changes in ideal gases

In this section the work and heat transfers for state changes from an initial state 'i' to a final state 'f' along different paths in closed systems are discussed. These state changes are typically associated with one constant thermodynamic variable and are named accordingly. State changes are typically visualized in TS- or pV-diagrams, where the heat Q and work W transferred during the state change can be directly read out.

Isothermal process: According to equation (A.6) the internal energy is constant for isothermal state changes, i.e. when $T = \text{const.}$ holds during the state change. Together with the first law of thermodynamics, equation (2.1), and the ideal gas equation (A.5) this leads to

$$Q_{T=\text{const.}} = -W_{T=\text{const.}} = \int_i^f pdV = nRT\int_i^f 1/VdV = nRT\ln\left(\frac{V_f}{V_i}\right). \tag{A.12}$$

In the TS-diagram isothermal state changes are parallel to the entropy axis, and in the pV-diagram they are given by the ideal gas equation equation (A.5):

$$p(V)_{T=\text{const}} = \frac{nRT}{V} \propto \frac{1}{V}. \tag{A.13}$$

The TS- and pV-diagrams for an isothermal expansion are shown in Figure A.1.

Isobaric process: In an isobaric state change the pressure remains constant. Here, first an expression for the work transferred and then, together with the first law of thermodynamics, also an expression for the heat tansferred during the process is given:

$$\begin{aligned} W_{p=\text{const.}} &= -p(V_f - V_i) = -nR(T_f - T_i) \\ Q_{p=\text{const.}} &= \int_i^f c_p(T)dT = nR\left(\frac{f(T_f)+2}{2}T_f - \frac{f(T_i)+2}{2}T_i\right) = H_f - H_i. \end{aligned} \tag{A.14}$$

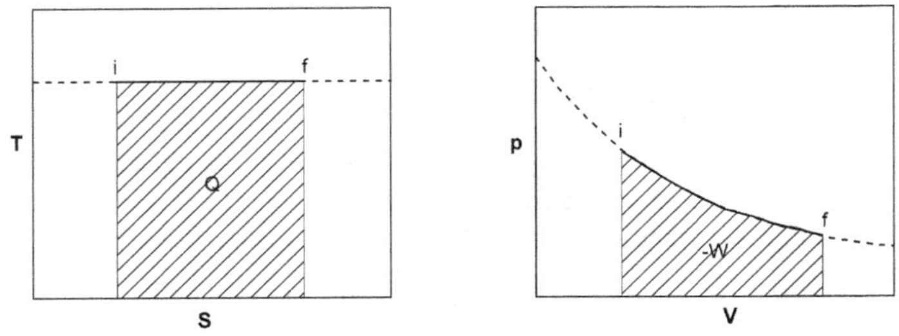

Fig. A.1: TS- and pV-diagram of an isothermal expansion of an ideal gas. The heat Q and work W are given by the marked areas.

In the pV-diagram an isobaric state change is a line parallel to the volume axis. For the curve in the TS-diagram the first law of thermodynamics (equation (2.1)) has to be taken into account, which together with the definition of the isobaric heat capacity given in equation (A.10) yields:

$$\begin{aligned} dU &= TdS - pdV \\ \Rightarrow dS &= \frac{dU + pdV}{T} = \frac{d(U+pV)}{T} = c_p \frac{dT}{T} \\ \Rightarrow S - S_i &= c_p \ln\left(\frac{T}{T_i}\right) \\ \Rightarrow T(S)_{p=\text{const}} &= T_i \cdot e^{(S-S_i)/c_p} \propto e^{S/c_p} \end{aligned} \tag{A.15}$$

The TS- and pV-diagrams for an isobaric expansion are shown in Figure A.2.

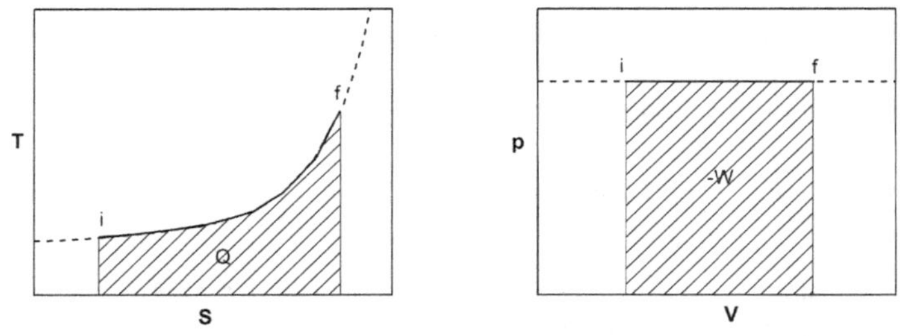

Fig. A.2: TS- and pV-diagrams of an isobaric expansion of an ideal gas. The heat Q and work W are given by the marked areas.

Isochoric process: The work transferred during an isochoric state change, i.e. a state change with $V = \text{const.}$, is zero. The heat transferred is then found to be

$$Q_{V=\text{const.}} = \int_i^f c_V(T)dT = nR\left(\frac{f(T_f)}{2}T_f - \frac{f(T_i)}{2}T_i\right) = U_f - U_i. \tag{A.16}$$

In the pV-diagram the isochoric state change is represented by a line parallel to the pressure axis. To find the curve representing an isochoric state change in the TS-diagram, again the first law of thermodynamics equation (2.1) and the definition of the isochoric heat capacity equation (A.9) have to be considered:

$$\begin{aligned} dU &= TdS - pdV = TdS \\ \Rightarrow \quad dS &= \frac{dU}{T} = c_V \frac{dT}{T} \\ \Rightarrow \quad S - S_i &= c_V \ln\left(\frac{T}{T_i}\right) \\ \Rightarrow \quad T(S)_{V=\text{const}} &= T_i \cdot e^{(S-S_i)/c_V} \propto e^{S/c_V} \end{aligned} \tag{A.17}$$

The isochoric state change has a steeper slope in the TS-diagram than the corresponding isobaric state change. The TS- and pV-diagrams for an isochoric heating process are shown in Figure A.3.

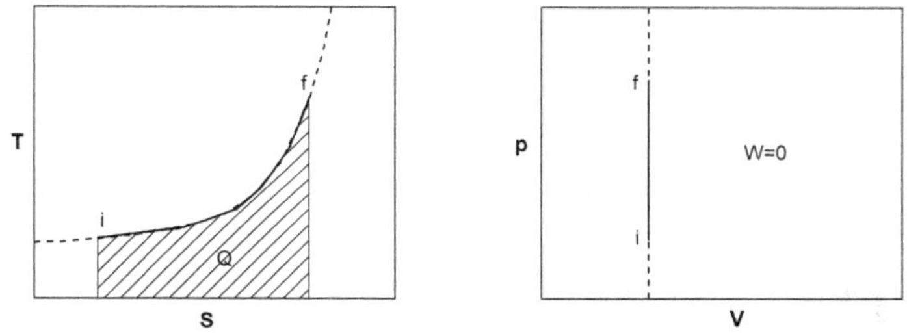

Fig. A.3: TS- and pV-diagrams of an isochoric heating of an ideal gas. The heat Q and work W are given by the marked areas.

Adiabatic (isentropic) process: An adiabatic state change is defined as a process where the transferred heat is zero. If the process is reversible, it is then also isentropic, i.e. the entropy stays constant. With the expression for the internal energy of an ideal gas given in equation (A.6), the first law of thermodynamics, equation (2.2), and the ideal gas equation (A.5), the following expression linking the change in pressure and volume is found:

$$\begin{aligned} -pdV &= dU = \frac{f}{2} nRdT = \frac{f}{2} d(pV) = \frac{f}{2}(Vdp + pdV) \\ \Rightarrow \quad -\left(\frac{f}{2} + 1\right)\frac{dV}{V} &= \frac{f}{2}\frac{dp}{p} \Leftrightarrow \frac{c_p}{c_V}\frac{dV}{V} = \frac{dp}{p} \\ \Rightarrow \quad \underbrace{\frac{c_p}{c_V}}_{:=\gamma} \ln\left(\frac{V_f}{V_i}\right) &= \ln\left(\frac{p_f}{p_i}\right) \\ \Rightarrow \quad \left(\frac{V_f}{V_i}\right)^{\gamma} &= \left(\frac{p_f}{p_i}\right) \Rightarrow p \propto \frac{1}{V^{\gamma}}, \end{aligned} \tag{A.18}$$

where γ is the adiabatic exponent. The TS- and pV-diagrams for an adiabatic expansion are shown in Figure A.4.

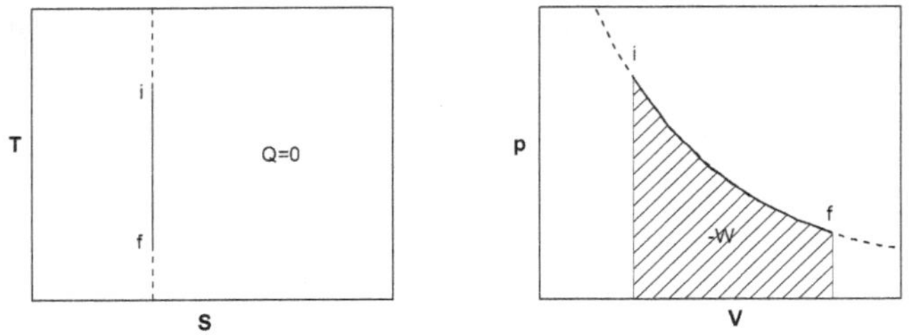

Fig. A.4: TS- and pV-diagrams of an adiabatic expansion of an ideal gas. The heat Q and work W are given by the marked areas.

The total work transferred during the adiabatic process can then be calculated using equation (A.18):

$$\begin{aligned} W &= -\int_i^f p(V)dV = -p_i V_i^\gamma \int_i^f V^{-\gamma} dV = -\left(\frac{p_i V_i^\gamma}{1-\gamma}\right)\left(V_f^{1-\gamma} - V_i^{1-\gamma}\right) \\ &= -\frac{f}{2} p_i V_i \left(\left(\frac{V_f}{V_i}\right)^{1-\gamma} - 1\right) = \frac{f}{2} nRT_i \left(\left(\frac{V_f}{V_i}\right)^{1-\gamma} - 1\right) \\ &= \frac{f}{2} nRT_i \left(\left(\frac{p_f}{p_i}\right)^{1-1/\gamma} - 1\right). \end{aligned} \tag{A.19}$$

Furthermore, an expression linking the temperatures and pressures or volumes can be found again using the equations (2.1), (A.5), and (A.6):

$$\begin{aligned} &\frac{f}{2} nRT_i \left(\left(\frac{p_f}{p_i} - 1\right)^{1-1/\gamma}\right) = W = \Delta U = \frac{f}{2} nR(T_f - T_i) \\ \Rightarrow\quad &\frac{T_f}{T_i} = \left(\left(\frac{p_f}{p_i}\right)^{1-1/\gamma}\right) \\ &\frac{T_f}{T_i} = \left(\left(\frac{V_f}{V_i}\right)^{1-\gamma}\right). \end{aligned} \tag{A.20}$$

A.3 The Carnot cycle

The second law of thermodynamics states that heat cannot be fully transformed into work. This means that a heat engine, i.e. an engine that performs work and is only driven by heat, can never reach a first law efficiency of $\eta_1 = 1$. An optimal heat engine follows the Carnot cycle, which therefore plays a special role among the thermodynamic cycles. Its TS- and pV- diagrams are depicted in Figure A.5.

For the following description of the individual state changes the working fluid is assumed to be an ideal gas. The heat and work transfers in the individual processes are calculated in the following:

1 ⟶ 2: In this adiabatic compression process no heat is transferred. The first law of thermodynamics readily determines the work transferred upon a state change from 1 to 2 denoted ${}_1W_2$:

$${}_1W_2 = U_2 - U_1. \tag{A.21}$$

The work is positive as the system has to be actively compressed.

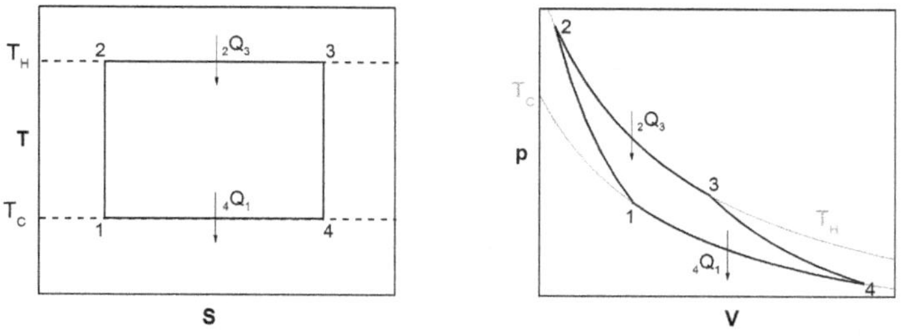

Fig. A.5: TS- and pV-diagrams of the Carnot cycle.

2 ⟶ 3: The system at temperature $T = T_h$ takes up the driving heat ${}_2Q_3$ in an isothermal process. For an ideal gas the internal energy remains constant, as it is proportional to the temperature as stated in equation (A.6). The first law of thermodynamics then yields

$$0 = \Delta U = {}_2Q_3 + {}_2W_3 \Rightarrow {}_2Q_3 = -{}_2W_3. \tag{A.22}$$

As ${}_2Q_3$ is positive the system expands providing work in this step (${}_2W_3 < 0$).

3 ⟶ 4: This step is again adiabatic and no heat is transferred. Analogous to 1 ⟶ 2 the work ${}_3W_4$ is transferred:

$$ {}_3W_4 = U_4 - U_3. \tag{A.23}$$

The system performs work while expanding adiabatically in this step, i.e. ${}_3W_4 < 0$.

4 ⟶ 1: In the final step, the system isothermally rejects the heat $|{}_4Q_1|$ to the heat bath at $T = T_c$. During this process the compression work $0 < {}_4W_1 = -{}_4Q_1$ has to be provided to the system.

As $U \propto T$ holds, the identities $U_1 = U_4$ and $U_2 = U_3$ also hold, with the consequence that

$$ {}_1W_2 + {}_3W_4 = U_2 - U_1 + U_4 - U_3 = 0. \tag{A.24}$$

This means that the work transferred across the borders of the system during both adiabatic processes cancels out. The overall work W performed by the system during the cycle can then be linked to the temperature levels during the heat tranfers in the following way:

$$\begin{aligned} W &= ({}_2W_3 + {}_4W_1) = -({}_2Q_3 + {}_4Q_1) = -(T_h(S_3 - S_2) + T_c(S_1 - S_4)) \\ &= -\Delta S\,(T_h - T_c)\,. \end{aligned} \tag{A.25}$$

Note that in the physical sign convention, already mentioned above in Section 2.1.1, work performed by the system is negative. Since efficiencies are positive they are calculated using absolute values. With this the overall efficiency of an Carnot engine becomes

$$\eta = \frac{|W|}{{}_2Q_3} = \frac{-({}_2W_3 + {}_4W_1)}{{}_2Q_3} = \frac{{}_2Q_3 + {}_4Q_1}{{}_2Q_3} = \frac{\Delta S(T_h - T_c)}{\Delta S \cdot T_h} = 1 - \frac{T_c}{T_h}. \tag{A.26}$$

A.4 Phase transitions

In a real gas the spatial extension of the molecules as well as the interaction between the individual molecules is taken into account. This is done assuming a strongly repulsive interaction between the particles for small radiuses and a quickly diminishing, attractive van der Waals interaction for larger particle distances. These assumptions modify the ideal gas equation and lead to the **van der Waals equation**:

$$\left(p + \frac{n^2 a}{V^2}\right)(V - nb) = nRT. \tag{A.27}$$

Here b accounts for the effective volume of the gas particles, and a describes a pressure correction due to the attractive interparticle interaction. The van der Waals equation offers a good description for real gases and their transition to the liquid phase. For real gases the internal energy is no longer proportional to the temperature. The phase of a substance is determined by the intensive variables pressure and temperature and can therefore be well visualized in pT-diagrams, an example of which is shown in Figure A.6.

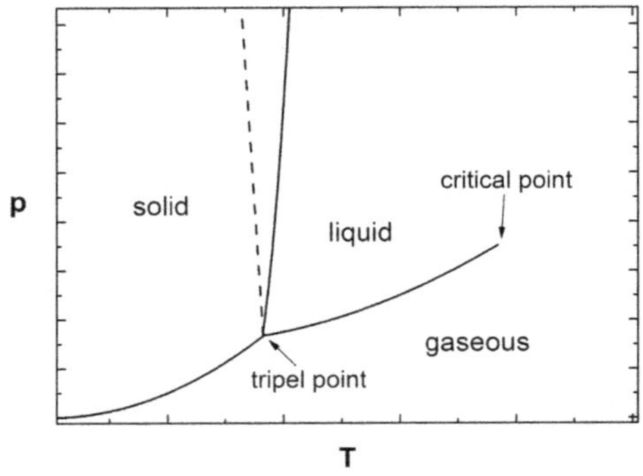

Fig. A.6: Example of a pT-diagram. The anomally of the solid/liquid separation curve for water is also sketched (*dashed line*).

The phase transitions to the solid phase are of minor importance in this book, as they do not occur in the systems considered. The liquid and the gaseous phase are only well distinguishable below a critical temperature $T = T_c$ and a critical pressure $p = p_c$. If one of the values is exceeded, a gradual change rather than a sharp phase transition in the behaviour is found. In this case the real gas is in the **supercritcal region**. This behaviour can be explained using the van der Waals equation (A.27), which offers a good description of systems in both the gaseous and the liquid phase close to the phase transition. Furthermore, it can be used to identify the location of the phase change between the liquid and the gaseous phase at temperatures $T < T_c$ using the so-called **Maxwell construction** as shown in Figure A.7.

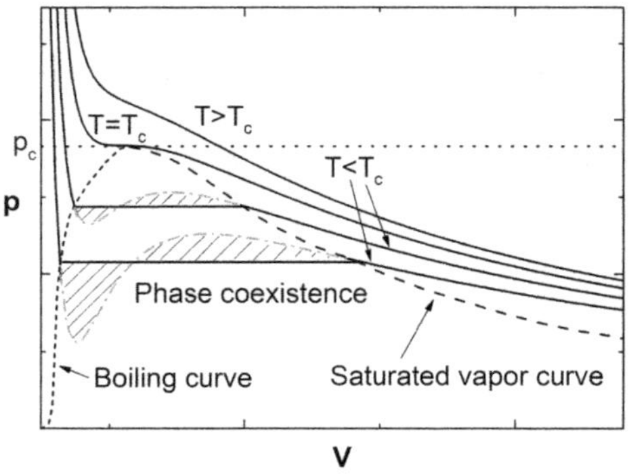

Fig. A.7: Four isotherms calculated using the van der Waals equation (A.27). For $T < T_c$ the nonphysical s-shape (*gray*) is circumvented by a flat characteristic that keeps the overall volume integral over the phase transition region constant. The two marked areas for each isotherm have to be chosen identical. The boiling curve (*short-dashed curve*, increasing) and saturated vapor curve (*dashed curve*, decreasing) are also shown.

The basis of this construction is the fact that the isotherms at $T < T_c$ given by the van der Waals equation show an unphysical behaviour, namely the increase in volume together with an increase of the pressure. The physical system circumvents this by adjusting a state in which both phases coexist, thereby minimizing the Gibbs free energy. During this process the pressure on the isotherms stays constant. The discontinuous points of the derivatives of the set of isotherms at $T < T_c$ mark the boiling curve (left) and the saturated vapor curve (right). Both curves meet in the critical point at the critical pressure p_c. In any region where more than one phase exists the number of intensive variables decreases according to the **Gibbs phase rule**:

$$F = K - P + 2. \tag{A.28}$$

Here F denotes the number of free (intensive) variables, K stands for the number of different substances, and P is the number of coexisting phases. According to equation (A.28) in the region of coexisting phases for a typical working fluid like water only one intensive variable, pressure, or temperature is required for the description of the system. Furthermore, the steam mass fraction x is defined in the following way:

$$x := \frac{m_g}{m_g + m_l}, \tag{A.29}$$

with the mass of gaseous substance m_g and liquid substance m_l. The specific heat required to evaporate a unit mass of the medium, i.e. changing x from 0 to 1, is called latent heat q_{lat}.

B. Basics: Solid state physics

In this chapter a short overview of the basic concepts from solid state physics relevant to this book are given. Some basic assumptions and relations will not be derived but simply given. The concepts underlying important relations relevant to this book, however, are derived from these basics.

B.1 Particle ensembles

To study the properties of particle ensembles in solid state, the quantum mechanical nature of the particles and the restrictions imposed by the solid state background have to be taken into account. In quantum mechanics there is always a collection of states which can either be occupied by particles or left unoccupied. The states exist, however, independently of their degree of occupation, and their properties are determined by the solid in which they exist. Properties of the particle ensemble are then, in turn, determined by both the distribution of states and the distribution of particles to the states. In the following three sections the distribution of available states in space, the distribution of the particles to these states, and the derivation of ensemble properties from the former two will be discussed.

B.1.1 Density of states

The density of states $D(E)dE$ gives the number of available particle states with energies in the interval $[E, E + dE]$ per unit volume. It depends on the geometry, dimensionality and, most importantly, on the dispersion relations $E(k)$ of the particles considered, where k is the wavenumber. In the case of electrons the periodicity of the crystal lattice imposes a structure on the density of states, where a nonzero density of states only occurs in discrete energy bands. For photons it is a continuous function. In this book only 3-dimensional problems are considered. The density of states can then be calculated in the following way:

$$D(E') \propto \int \delta(E(k) - E') \frac{d^3k}{(2\pi)^3} =: \frac{1}{(2\pi)^3} \int \frac{1}{dE/dk} dS \propto \int \frac{1}{v_g} dS, \tag{B.1}$$

where δ is the δ-distribution, and S stands for the area in reciprocal space perpendicular to the k-vectors at the given energy. As often in solid states physics, the **reciprocal space** is used for the calculation of the density of states. It is also called **k-space** and is the spatial Fourier transform of normal space. The density of states is inversely proportional to the group velocity v_g of the particles at a given point. If a spatially uniform solid is assumed, the following simplification, corresponding to a spherical geometry in the reciprocal space, is valid:

$$\frac{1}{(2\pi)^3} \int dS \to \frac{1}{(2\pi)^3} 4\pi k_0^2 = \frac{k'^2}{2\pi^2}. \tag{B.2}$$

With this, the density of states for the two types of dispersion relations considered in this book, one for photons and one for electrons close to a band edge, can be calculated. The two dispersion relations for these types of particles read

$$\begin{aligned} &\text{photons:} && E(k) = \hbar c k \\ &\text{electrons close to band edge:} && E(k) = E_{\text{edge}} \pm \frac{(\hbar k)^2}{2m_{\text{eff}}}, \end{aligned} \tag{B.3}$$

where m_{eff} stands for the **effective mass** of the electrons and the '+' refers to electrons at a lower and '−' to electrons at an upper band edge. Noting that the crystal momentum is given by $\hbar k$, the second term of the dispersion relation for electrons close to the band edge can thus be interpreted as the kinetic energy of free particles with mass m_{eff}. In case of an upper band edge this leads to the assumption of a negative effective mass and the effective particles are called **defect electrons** or **holes**.

In the case of photons the first proportionality factor in equation (B.1) is 2, reflecting the two possible polarization directions. The density of states then becomes

$$D_{\text{photon}}(E)dE = 2\frac{k^2}{2\pi^2\hbar c}dE = \frac{E^2}{\pi^2(\hbar c)^3}dE = \frac{8\pi E^2}{(hc)^3}dE. \tag{B.4}$$

For the electrons and holes close to the band edge the energy with respect to the band edge $\varepsilon := |E - E_{\text{edge}}|$ is relevant. The proportionality factor in equation (B.1) is equal to $2M_{\text{c,v}}$, the factor 2 this time reflecting the two possible spin states of the electrons/holes and a possible, material specific degeneracy $M_{\text{c,v}}$ of the energy extremum of conduction and valence band, respectively:

$$\begin{aligned} D_{\text{e}}(\varepsilon)d\varepsilon &= 2M_{\text{c}}\frac{m_{\text{eff,e}}k}{2\pi^2\hbar^2}d\varepsilon = M_{\text{c}}\frac{\sqrt{2}m_{\text{eff,e}}^{3/2}}{\pi^2\hbar^3}\sqrt{\varepsilon}d\varepsilon \\ D_{\text{h}}(\varepsilon)d\varepsilon &= 2M_{\text{v}}\frac{m_{\text{eff,h}}k}{2\pi^2\hbar^2}d\varepsilon = M_{\text{v}}\frac{\sqrt{2}m_{\text{eff,h}}^{3/2}}{\pi^2\hbar^3}\sqrt{\varepsilon}d\varepsilon \end{aligned} \tag{B.5}$$

Note that the degenerate states can also be associated with different effective masses, in which case an averaged effective mass has to be used. This means that close to the band edge the density of states increases with the square root of the energetic distance to the band edge. Particles with higher absolute values of the effective mass m_{eff} show a steeper increase in the density of states.

B.1.2 Distribution functions and the electrochemical potential

After the calculation of the density of states in the last section, the next task is to determine with which probability states at a given energy are occupied. This is first of all a thermodynamic question. As always the equilibrium state is characterized by a minimized free energy, in this case G. In this equilibrium state, characterized by temperature T, pressure p, and chemical potential μ of the particles under consideration, the free energy G_{E} of a state at energy E is given by

$$G_{\text{E}} = -kT \cdot \sum_{j=0}^{\infty} \ln\left(e^{(\mu-E)/kT}\right)^j, \tag{B.6}$$

where j is the number of particles occupying the state given. There are two main classes of particles: **fermions** and **bosons**. Examples of fermions are electrons, protons, neutrons, and also holes and examples for bosons are photons and phonons. Fermions are subject to the Pauli exclusion principle, which means that no two particles can occupy the same state, i.e. $j \in \{0, 1\}$, whereas the bosons are not subject to this restriction. For bosons, G_{E} is a geometrical series, and a finite value is thus only given if for the chemical potential $\mu \leq 0$ holds. These differences lead to different free energies for the two particle types:

$$\begin{aligned} G_{\text{E}} &= -kT \cdot \ln\left(1 + e^{(\mu-E)/kT}\right): && \text{fermions,} \\ G_{\text{E}} &= -kT \cdot \ln\left(\frac{1}{1 - e^{(\mu-E)/kT}}\right) = kT \cdot \ln\left(1 - e^{(\mu-E)/kT}\right): && \text{bosons.} \end{aligned} \tag{B.7}$$

The time average of the particle number $f(E,T)$ in a state with energy E is then given by $f(E,T) = -(\partial G_E/\partial\mu)_{p,T}$, resulting in the Fermi–Dirac (fermions) and Bose–Einstein (bosons) distributions:

$$\begin{aligned} f(E,T) &= \frac{1}{e^{(E-\mu)/kT}+1}: \quad \text{fermions} \\ f(E,T) &= \frac{1}{e^{(E-\mu)/kT}-1}: \quad \text{bosons.} \end{aligned} \tag{B.8}$$

In the case of fermions the Fermi–Dirac distribution $f(E,T)$ can be interpreted as the occupation probability of the states at energy E. In the course of this book only one special case for each particle class is important: electrons in solids and photons from black body radiation. In both cases the general distribution functions are subject to small case-specific modifications. For electrons in solids the fact that the particles are charged has to be considered. The energy required to add or subtract a particle from the ensemble is thus dependent on the electrostatic potential φ at the point where the particle transfer takes place. This can be captured by replacing the chemical potential in equation (B.8) with the electrochemical potential $\tilde{\mu} = \mu - e\varphi$.

Photons in thermal equilibrium with matter represent a special case where the number of particles is not constant. This means that no energy is gained by adding or subtracting a photon from the ensemble. This corresponds to $\mu = 0$. The relevant distribution functions for the two cases thus read

$$\begin{aligned} f(E,T) &= \frac{1}{e^{(E-\tilde{\mu})/kT}+1}: \quad \text{electrons,} \\ f(E,T) &= \frac{1}{e^{E/kT}-1}: \quad \text{photons.} \end{aligned} \tag{B.9}$$

In Figure B.1 the distribution functions for electrons in a solid with $\tilde{\mu} = 4$ eV and for photons are shown for different temperatures.

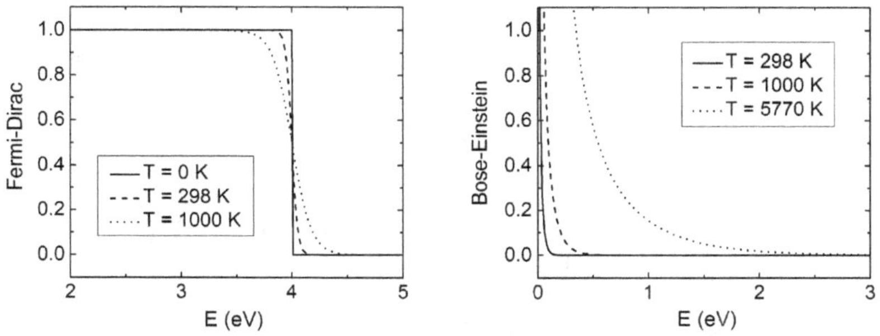

Fig. B.1: Fermi–Dirac distributions for electrons with $\tilde{\mu} = 4$ eV (*left*) and Bose-Einstein distributions for photons (*right*) at different temperatures.

B.1.3 Selected properties

Important properties of the particle ensembles as a whole, as, e.g. the volume density n or the volume specific internal energy u, can now be calculated in the following way, using the density of states and the appropriate distribution function (see B.1.2):

$$\begin{aligned} n &= \int_0^\infty D(E)f(E,T)dE =: \int_0^\infty \frac{dn}{dE}dE \\ u &= \int_0^\infty D(E)f(E,T)EdE =: \int_0^\infty \frac{du}{dE}dE, \end{aligned} \tag{B.10}$$

where the terms dn/dE and du/dE are the **particle density** and **energy density spectrum**, respectively.

In a photon ensemble the free energy per unit volume $g = G/V$ can be calculated using the free energy per state as given in equation (B.6):

$$g = \int_0^\infty D(E)kT \cdot \ln\left(1 - e^{-E/kT}\right) dE = kT\frac{8\pi}{h^3c^3}\int_0^\infty E^2 \ln\left(1 - e^{-E/kT}\right) dE. \tag{B.11}$$

With this the volume specific entropy is also found via $s = -(\partial g/\partial T)_p$:

$$\begin{aligned} s &= -\frac{8\pi}{(hc)^3}\left(kT\int_0^\infty E^2 \cdot \left(\frac{1}{1-e^{-E/kT}}\right)\cdot\left(-e^{-E/kT}\right)\cdot\left(\frac{E}{(kT)^2}\right)dE \right. \\ &\qquad \left. + k\int_0^\infty E^2\ln\left(1-e^{-E/kT}\right)dE\right) \\ &= \frac{8\pi}{(hc)^3}\left(\frac{1}{T}\int_0^\infty \frac{E^3}{e^{E/kT}-1}dE - k\int_0^\infty E^2 \ln\left(1-e^{-E/kT}\right)dE\right) \\ &= \frac{1}{T}u - k\int_0^\infty E^2\ln\left(1-e^{-E/kT}\right)dE. \end{aligned} \tag{B.12}$$

The last identity makes use of equation (B.10). With partial integration the integrals can be solved:

$$\begin{aligned} s &= \frac{8\pi}{(hc)^3}\left(\frac{1}{T}\int_0^\infty \frac{E^3}{e^{E/kT}-1}dE - k\int_0^\infty E^2\ln\left(1-e^{-E/kT}\right)dE\right) \underbrace{=}_{y:=E/kT} \\ &= \frac{8\pi}{(hc)^3}k^4T^3\left(\int_0^\infty \frac{y^3}{e^y-1}dy - \left(\underbrace{\frac{1}{3}y^3\ln\left(1-e^{-y}\right)\Big|_0^\infty}_{=0} - \frac{1}{3}\int_0^\infty \frac{y^3}{e^y-1}dy\right)\right) \\ &= \frac{4}{3}\frac{8\pi}{(hc)^3}k^4T^3\underbrace{\int_0^\infty \frac{y^3}{e^y-1}dy}_{=\pi^4/15} = \frac{32\pi^5}{45(hc)^3}k^4T^3 =: \frac{16}{3c}\sigma T^3, \end{aligned} \tag{B.13}$$

where $\sigma = 2\pi^5k^4/(15h^3c^2)$ is the Stefan–Boltzmann constant. It is also possible to calculate the volume specific entropy of a boson ensemble directly from equation (B.11) by integrating the free energy first and then taking the derivative of the result. In this case, however, the information on the spectral distribution of the entropy as laid out in equation (B.12) is lost.

B.1.4 The Sommerfeld expansion

The Sommerfeld expansion is a mathematical expansion for properties of ideal fermion ensembles. Such properties, denoted here with P, can in general be calculated via equation (B.10), i.e. as an energy integral over a function $g(E) = \tilde{g}(E) \cdot D(E)$ multiplied with the Fermi–Dirac distribution. It is based on the assumption of a constant particle number and valid for low temperatures, where $kT \ll \tilde{\mu}$ holds. The Sommerfeld expansion makes use of the fact that the Fermi–Dirac distribution only changes significantly very close to $E = \tilde{\mu}$ and that all important properties are thus only determined by a small energy region around $\tilde{\mu} \approx E_F$. In this region the function $g(E)$ then changes only slightly. All functions are set to zero for negative energies and the antiderivative of $g(E)$ is given by

$$G(E) := \int_{-\infty}^{E} g(E')dE'. \tag{B.14}$$

With this the following exact expression for P can be derived:

$$\begin{aligned} P &= \int_0^\infty f(E,T)g(E)dE = \int_{-\infty}^{\infty} f(E,T)g(E)dE \\ &= f(E,T)g(E)\Big|_{-\infty}^{\infty} - \int_{-\infty}^{\infty} \frac{df(E,T)}{dE} G(E)dE \\ &= -\int_{-\infty}^{\infty} \frac{df(E,T)}{dE} G(E)dE = \int_{-\infty}^{\infty} \frac{1}{kT} \frac{e^{(E-\tilde{\mu})/kT}}{\left(e^{(E-\tilde{\mu})/kT}+1\right)^2} G(E)dE \\ &\underbrace{=}_{x:=(E-\tilde{\mu})/kT} \int_{-\infty}^{\infty} \frac{e^x}{(e^x+1)^2} G(\tilde{\mu}+xkT)dx, \end{aligned} \tag{B.15}$$

where the first term in the second line is equal to zero as both functions are zero at the lower boundary and $f(E,T)$ is zero for the upper boundary while $g(E)$ is at least finite there. As the derivative of the Fermi–Dirac distribution only changes significantly around $E = \tilde{\mu} \approx E_F$, the antiderivative $G(\tilde{\mu} + xkT)$ on the right-hand side of equation (B.15) can be expanded into a Taylor series:

$$G(\tilde{\mu} + xkT) = G(\tilde{\mu}) + xkTG'(\tilde{\mu}) + \frac{1}{2}(xkT)^2 G''(\tilde{\mu}) + o(T^3). \tag{B.16}$$

With this expansion the exact value of P can be written as follows:

$$\begin{aligned} P &= \sum_{k=0}^{\infty} \frac{(kT)^k}{k!} G^{(k)}(\tilde{\mu}) \int_{-\infty}^{\infty} \frac{x^k e^x}{(e^x+1)^2} dx \\ &= \sum_{k=0}^{\infty} \frac{(kT)^{2k}}{(2k)!} G^{(2k)}(\tilde{\mu}) \int_{-\infty}^{\infty} \frac{x^{2k} e^x}{(e^x+1)^2} dx, \end{aligned} \tag{B.17}$$

where the last identity holds as $e^x/(e^x+1)^2 = (e^{(x/2)} + e^{-(x/2)})^{-2}$ is an even function. Integrals as on the right hand side of equation (B.17) can be exactly calculated and the Sommerfeld expansion up to terms $o(T^2)$ thus takes the following form:

$$\begin{aligned} P &\approx G(\tilde{\mu}) \underbrace{\int_{-\infty}^{\infty} \frac{e^x}{(e^x+1)^2} dx}_{=1} + \frac{1}{2}(kT)^2 G''(\tilde{\mu}) \underbrace{\int_{-\infty}^{\infty} \frac{x^2 e^x}{(e^x+1)^2} dx}_{=\pi^2/3} \\ &= G(\tilde{\mu}) + \frac{\pi^2}{6}(kT)^2 g'(\tilde{\mu}). \end{aligned} \tag{B.18}$$

With this expansion the temperature dependence of the electrochemical potential can be calculated by setting $g(E) \rightarrow D(E)$, i.e. as the density of states. Note that again the assumption is always a constant particle density:

$$\begin{aligned}
N &= \int_0^\infty f(E,T)D(E)dE = \int_{-\infty}^\infty f(E,T)D(E)dE \approx \int_{-\infty}^{\tilde{\mu}} D(E)dE + \frac{\pi^2}{6}(kT)^2 D'(\tilde{\mu}) = \\
&\approx \underbrace{\int_{-\infty}^{E_F} D(E)dE}_{=N} + (\tilde{\mu} - E_F)D(\tilde{\mu}) + \frac{\pi^2}{6}(kT)^2 D'(\tilde{\mu}) \qquad \text{(B.19)} \\
\Rightarrow \quad \tilde{\mu} &\approx E_F - \frac{\pi^2}{6}(kT)^2 \frac{D'(\tilde{\mu})}{D(\tilde{\mu})},
\end{aligned}$$

where $D'(\tilde{\mu})$ is the derivative with respect to E of the density of states at the electrochemical potential. Note that the Sommerfeld expansion, while providing a very good approximation in metals, fails completely in semiconductors. The reason for this is that the functions $g(E)$ that are usually linked to the density of states $D(E)$ are not varying slowly in the region where the Fermi–Dirac distribution changes.

B.2 Semiconductors

In this book semiconductors play a special role among the solids, because they are used in photovoltaics. For this reason the electronic structure and optical properties of different kinds of semiconductors are discussed in this section. In any crystalline bulk material the dispersion relation of the electrons is degenerate. The reason for this is the periodicity of the material and the consequent periodicity of the crystal momentum. For any crystal momentum there are as many distinct energy levels as unit cells in the crystal. These distinct energy states are numbered with the so-called band index, and all states with the same band index form an energy band. These energy bands are then filled with the electrons in the crystal starting with the lowest energy obeying the Fermi–Dirac distribution as described in equation (B.9). The band with the highest index which still contains electrons is called the **valence band**, in analogy to the outermost shell in atoms. If the valence band is partly filled, the solid is a conductor, because the electrons in the valence band can be continuously excited by electric fields. To excite an electron in the case of a completely filled valence band, the energy difference to the next unoccupied state has to be overcome. As this state is in the band with the next higher index, which is called **conduction band**, the energy gap, or **bandgap** E_g, between both bands has to be overcome to perform the excitation. In this case the material is classified as an insulator ($E_g \geq 3$ eV) or semiconductor ($E_g \leq 3$ eV), depending on the energy gap between the bands.

B.2.1 Intrinsic semiconductors

The conductivity σ of intrinsic semiconductors is many orders of magnitude lower than the conductivity of metals. The reason for this is the low density of mobile charge carriers n_i in any partially filled band. All mobile charge carriers have to be thermally excited, and as every excited electron in the conduction band leaves one hole, i.e. unoccupied electron states close to the band edge behaving like charge carriers with a positive charge, in the valence band the number of mobile electrons n_e and of mobile, positively charged holes n_h is always identical:

$$n_i = n_e = n_h. \tag{B.20}$$

The charge carrier density in the respective bands can be calculated using equation (B.10). For the calculation, shown for the conduction band as an example, the following condition, corresponding to a **nondegenerate** semiconductor is assumed:

$$\tilde{\mu} - E_{VB} > 2kT \wedge E_{CB} - \tilde{\mu} > 2kT \quad \Rightarrow \quad e^{(E-\tilde{\mu})/kT} + 1 \approx e^{(E-\tilde{\mu})/kT}, \tag{B.21}$$

where E_{VB} and E_{CB} are the energy of the valence and conduction band edges, respectively. Furthermore the following identity is used:

$$\int_0^\infty \sqrt{x}e^{-x}dx = \frac{\sqrt{\pi}}{2}. \tag{B.22}$$

Using equation (B.10) and the abbrevation $\varepsilon := E - E_{CB}$ the charge carrier density can be linked to the position of the electrochemical potential:

$$\begin{aligned}
n_e &= \int_0^\infty D_{CB}(\varepsilon) \cdot f(E,T)d\varepsilon \approx \int_0^\infty D_{CB}(\varepsilon) \cdot e^{-(\varepsilon+E_{CB}-\tilde{\mu})/kT} d\varepsilon \\
&= M_c \frac{\sqrt{2}m_{\text{eff}}^{3/2}}{\pi^2\hbar^3} \cdot e^{-(E_{CB}-\tilde{\mu})/kT} \int_0^\infty \sqrt{\varepsilon} \cdot e^{-\varepsilon/kT} d\varepsilon \\
&\underbrace{=}_{y:=\varepsilon/kT} M_c \frac{\sqrt{2}(kT \cdot m_{\text{eff}})^{3/2}}{\pi^2\hbar^3} \cdot e^{-(E_{CB}-\tilde{\mu})/kT} \int_0^\infty \sqrt{y} \cdot e^{-y} dy \\
&\underbrace{=}_{\text{eq. (B.22)}} \frac{M_c}{\sqrt{2}\hbar^3} \cdot \left(\frac{kT \cdot m_{\text{eff}}}{\pi}\right)^{3/2} \cdot e^{-(E_{CB}-\tilde{\mu})/kT} =: N_C \cdot e^{-(E_{CB}-\tilde{\mu})/kT},
\end{aligned} \tag{B.23}$$

where N_C is called the effective density of states in the conduction band. It is a doping-independent material quantity only dependent on the effective mass of the electrons and temperature. At $T = T_{am}$ it is of the order of $N_C \approx 10^{19}\ \text{cm}^{-3}$ for silicon. A similar calculation can be done for the valence band yielding

$$n_h = N_V \cdot e^{-(\tilde{\mu}-E_{VB})/kT}, \tag{B.24}$$

where N_V for the valence band plays the role of N_C for the conduction band. The intrinsic density of mobile charge carriers n_i can then be calculated using equation (B.20):

$$n_i = \sqrt{n_e n_h} = \sqrt{N_C N_V e^{-E_g/2kT}}. \tag{B.25}$$

For the position of the electrochemical potential the following is then valid according to equations (B.23) and (B.24):

$$\begin{aligned}
E_{CB} - \tilde{\mu} &= kT \cdot \ln\left(\frac{N_c}{n_e}\right) = kT \cdot \ln\left(\frac{N_c}{n_i}\right) \\
\tilde{\mu} - E_{VB} &= kT \cdot \ln\left(\frac{N_v}{n_h}\right) = kT \cdot \ln\left(\frac{N_v}{n_i}\right) \\
\Rightarrow \quad \tilde{\mu} &= \frac{E_{CB} + E_{VB}}{2} + \frac{1}{2}kT \cdot \ln\left(\frac{N_v}{N_c}\right).
\end{aligned} \tag{B.26}$$

This means that the Fermi energy, i.e. $\tilde{\mu}(T = 0)$, lies exactly in the middle of the bandgap. As the effective densities of states $N_{c,v}$ are typically of the same order of magnitude, and as kT is much smaller than $\tilde{\mu}$, the electrochemical potential is also close to the middle of the bandgap for nonzero temperatures.

B.2.2 Doping

An increase by several orders of magnitude of semiconductor conductivity can be achieved by doping of the semiconductor, i.e. by adding small amounts (ppb to ppm) of suitable impurities. These impurities can either be ionized by electrons from the valence band generating free holes in the valence band or by injecting electrons into the conduction band. The former doping, called p-type doping, can be achieved for example in silicon by replacing some silicon atoms with atoms of valence 3, such as e.g. boron or aluminum, so-called acceptors, and the latter, called n-type doping, by doing the same with atoms of valence 5, such as e.g. phosphor or arsenic, called donors. The energy levels of the impurities are close to the respective band edges (several 10 meV), which means that they can be easily excited by thermal energy. This leads to the so-called **impurity depletion** at room temperature, where all impurities are ionized thermally. In this case the expressions $n_e \approx n_D \gg n_i$ where n_D is the density of donor atoms for n-doped material, and $n_h \approx n_A \gg n_i$ where n_A is the density of acceptor atoms for p-doped material hold. Correspondingly, the electrochemical potential shifts to a value close to the band edge which can be calculated using equation (B.26):

$$\begin{aligned} E_{CB} - \tilde{\mu} &= kT \cdot \ln\left(\frac{N_C}{n_e}\right) \approx kT \cdot \ln\left(\frac{N_C}{n_D}\right) \\ \tilde{\mu} - E_{VB} &= kT \cdot \ln\left(\frac{N_V}{n_h}\right) \approx kT \cdot \ln\left(\frac{N_V}{n_A}\right). \end{aligned} \tag{B.27}$$

In a doped semiconductor the mobile charge carrier type generated by the doping is called a **majority charge carrier**. An example for the energetic distribution of holes in the valence band for a p-type semiconductor (in this case p-type doped silicon) is shown in Figure B.2.

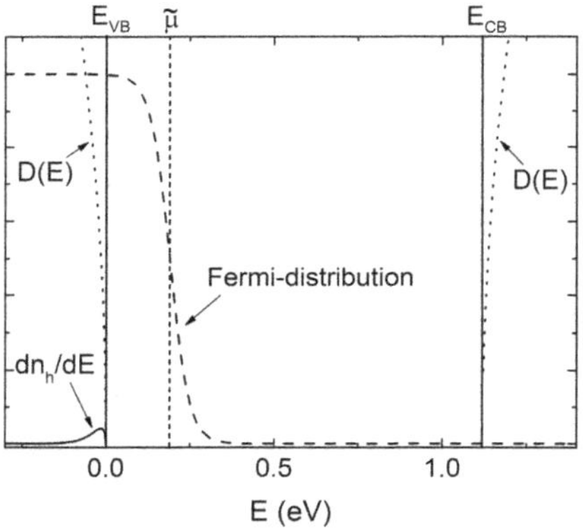

Fig. B.2: Energetic distribution of holes in the valence band (*solid curve*) of p-type silicon at $T = 60\,^\circ$C together with the Fermi–Dirac distribution (*dashed curve*) and the densities of states in the valence and the conduction band (*dotted curves*). The distribution of the holes is given by $dn_h/dE = D(E)(1 - f(E))$ (B.10), as they correspond to unoccupied electron states.

B.2.3 Optical properties

In an absorbing medium the intensity of light I is a function of the distance d from the surface and can be described by the **Lambert–Beer law.**

$$I(d) = I_0 \cdot e^{-\alpha d}. \tag{B.28}$$

Here α stands for the **absorption coefficient**, which is typically dependent on the photon energy, i.e. $\alpha \rightarrow \alpha(h\nu)$. This phenomenon is most prominent in semiconductors where $\alpha(h\nu)$ sharply changes

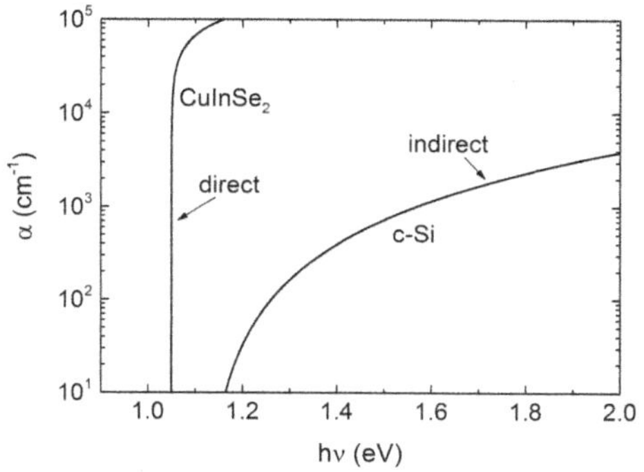

Fig. B.3: Dependence of the absorption coefficient on the photon energy $\alpha(E_{ph})$ for crystalline silicon (c-Si) and $CuInSe_2$. The former material is an indirect, and the latter a direct semiconductor.

over several orders of magnitude from essentially no absorption for photons with an energy smaller than the band gap of the semiconductor to a significant absorption for photons with energies above this threshold. This behaviour can be understood by taking into account that photons interact with the electrons of the absorbing material. If the photon energy is below the band gap, it cannot excite the electrons from the inert valence band, as there are no final states for the electrons available inside the band gap. For photons with a higher energy than the band gap, interband transitions become possible, and the empty conduction band provides the necessary final states for the electron excitation. In these interband excitation processes e^-/h^+-pairs are generated as the electron excited into the conduction band leaves a hole in the valence band. The value for the absorption coefficient depends on whether the semiconductor has a direct or an indirect bandgap, where in the latter case phonons additional to photons are necessary for the e^-/h^+-pair formation. This is due to the fact that the photons alone do not carry the momentum necessary to bridge the difference in crystal momentum between the ground state of the electron in the valence band and the final electronic state in the conduction band. The additional momentum has to be provided by a suitable phonon, which makes the excitation process a three-particle process, thus reducing its probability. As a result in direct semiconductors the absorption coefficients are in general higher and have a steeper dependence on $\varepsilon = h\nu - E_g$. As an example, in Figure B.3 the energy dependence of the absorption coefficient for the two most widely used semiconductors in photovoltaics, crystaline silicon (indirect) and $CuInSe_2$ (direct), are shown. Due to the much stronger absorption, thin film solar cells with absorber thicknesses of only several μm are always built from direct semiconductors.

C. Further Reading

For all readers interested in a more in-depth study of the topics described in this book, a small list of literature suggestions is given here.

A modern introduction to thermodynamics, which is relevant for all chapters of this book can be found in D. Kondepudi and I.Prigogine, *Modern Thermodynamics, from Heat Engines to Dissipative Structures*, John Wiley & Sons, Chichester, 1999.

Concepts from thermodynamics and solid state physics are treated from a statistical point of view in L. D. Landau and E. M. Lifshitz, *Statistical Physics*, Course of theoretical physics 5, Elsevier, 1980.

A classical textbook on solid state physics containing a clear and comprehensive introduction into the topic is N. W. Ashcroft and N. D. Mermin, *Solid State Physics*, Saunders College, Philadelphia, 1976.

Application of thermodynamics to a broad range of engineering applications, including steam and gas power plants is given in M. J. Moran, H. N. Shapiro, D. D. Boettner, and M. B. Bailey, *Fundamentals of Engineering Thermondynamics*, 7th ed., John Wiley & Sons, New York, 2010.

An in-depth discussion of the physical concepts behind solar cells, including a concise introduction to solid state physics, as well as a survey of different solar cell realizations and advanced concepts in solar cell development can be found in P. Würfel, *Physics of Solar Cells: From Basic Principles to Advanced Concepts*, Wiley-VCH, Weinheim, 2009.

The book J. larminie and A. Dicks, *Fuel Cell Systems Explained*, John Wiley & Sons, Chichester, 2003 gives an introduction to fuel cell technology, covering low, middle, and high temperature fuel cells as well as the fueling of fuel cells and system aspects.

Index

Abbe sine condition 137
absorption
- coefficient 194
activation barrier 113, 114
activity 91
- dissolved species 93
- gases 93
- liquids 93
- solids 93
adiabatic exponent 64, 179
affinity 91
air mass AM 125
- AM0 125
- AM1 125
- AM1.5 125
albedo 125
ambient state 11
anergy 11
anergy transfer
- closed system 24
- open system 38
anode 99
atmosphere 124
ATP/ADP 128

back work ratio 65
band structure 192
bandgap 192
battery
- alkaline 116
- capacity 117
- Li-ion 174
- nonrechargeable 116
- rechargeable 174
bipolar plates 116
boiling curve 184
Bose-Einstein distribution 119, 189
boson 188
built-in asymmetry 79
- electrochemical 106
- photovoltaic 143
Butler-Volmer equation 114, 158

Carnot, Sadi 8
Carnot cycle 8, 23, 182
carnotization 59
cathode 99
chemical potential 74
- molar 75
chlorophyll 128
chloroplasts 127
Clausius, Rudolf 8, 18
CO_2 emission 45, 127
coefficient of fugacity 92
combustion
- hydrogen 106
- methane 87, 89, 90
- second-law efficiency 97, 98
combustion chamber 63
complex impedance 80
compressed air energy storage CAES 167, 168
compressor 62
- CAES 168
- gas turbine 61, 62
- heat pump 28
- piston 35
concentrating solarthermal energy 136, 141
- second-law efficiency 140
concentration factor 138
condenser 48
- second-law efficiency 56
conduction band 192
contact voltage 76
control volume 31

dark reaction 128
dark saturation current 158
defect electrons 188
density of states 146, 187
- photon ensemble 188
- semiconductor 188
detailed balance limit 154
diffusion current 77
dispersion relation 187
distribution
- Bose-Einstein 119, 189
- electrons 189
- Fermi–Dirac 73

– Fermi-Dirac 146, 189
– photons 189
double layer 76, 101
– capacitance 102
drift current 77
drift velocity 77
driving force 13
– chemical reaction 90
– current 77

effective density of states 146, 193
effective mass 188
efficiency
– absorption 153
– Betz 43
– Carnot 9
– first-law 12
– radiative recombination 151
– second-law 13
– thermal 109
electrical
– current 77
– field 77
– generator, first-law efficiency 82
electrochemical potential 73, 189
– electrons in electrolyte 103
– molar 75
electrochemistry 99
electrode potential 101
– standard 105
electrolytic cell 173
– UI-characteristic 173
electron ensemble 73
electron/hole-pair 143
electrostatic potential 74
energy
– final 1
– net 1
– primary 1
– productivity 7
– renewable 1
– sustainable 1
energy consumption 6
– EU 4, 8
– USA 4, 8
energy flow diagram 5
– combined cycle power plant 71
– direct electrical heating 13
– EU 8
– gas turbine power plant 68
– heat pump 13, 27
– solar cell 162
– steam power plant 54
– USA 8
enthalpy 16, 177
– formation 88
– reaction 88
– standard reaction 88
entropy
– heat radiation 190
– mixture 92
– standard 89
– standard reaction 89, 109
equilibrium cell voltage 104
equipartition theorem 178
exchange current density 114
exergy 11
– chemical 87
– closed system 19, 21
– fuels 94
– open system 37
– sources 25
exergy flow diagram 12
– combined cycle power plant 71
– gas turbine power plant 68
– heat pump 27
– solar cell 162
– steam power plant 54
exergy function
– closed system 21
– open system 37
exergy storage 167
– electrochemical 172
– mechanical storage 167
– thermal 169
exergy storage chemical 172
exergy transfer 24
– via electron transfer 75
– via isobaric heat transfer 39
– via isothermal heat transfer 22, 38
– via work 22, 38
extent of reaction 90

Faraday constant 75
feed water heaters
– closed 59
– open 59
feed water pump 47

Fermi energy 74
fermion 188
figure of merit 85
filling factor 160
first law of thermodynamics
– closed system 16
– open system 32, 34
forward bias 157
fossil fuels 2
Fresnel lens 142
fuel cell 115
– H_2/O_2 116

galvanic cell 108
– IU-characteristic 112
– ohmic losses 112
– second-law efficiency 111
– UI-characteristic 113
gas turbine power plant
– closed 62
– first-law efficiency 64, 66–68
– open 60, 61
– second-law efficiency 67, 69
generator 45, 79
Gibbs free energy 16, 177
– quantum mechanical ensemble 188
– reaction 94
– standard reaction 89
Gibbs phase rule 185
Gouy–Chapman layer 102
greenhouse effect 126
– anthropogenic 127
group velocity 187

heat 16
heat exchanger 38
– counterflow 39
– second-law efficiency 39, 40
heat loss coefficient 133
heat pump 13, 26, 172
– absorption 26
heat radiation 133, 151
heating value 2, 88
– upper 2, 88
heliostat 142
Helmholtz free energy 16, 177
Helmholtz layer 102
high temperature heat storage 141, 170
holes 188
hydrogen combustion reaction 106
– half cell reactions 116

ideal gas 178
– adiabatic exponent 179
– equation 178
– heat capacity 178
impurity depletion 194
intercalation 174
intercooling 69
interface 75
– electrode/electrolyte 99
Intergovernmental panel on climate change (IPCC) 127
internal energy 16, 177
– ideal gas 178
intrinsic charge carrier density 150, 192

Joule–Brayton cycle 62
Joule-Brayton cycle 63

k-space 187

Lambert–Beer law 194
latent heat 185
law of induction 79
light harvesting complex (LHC) 129
light reaction 127
low temperature heat storage 171

magnetic field 80
magnetic flux 79
majority charge carrier 155, 194
maximum power point MPP 160
Maxwell construction 184
membrane electrode assembly (MEA) 117
methane combustion 87, 89, 90
minority charge carrier 155
– diffusion length 158
– lifetime 158
mobility 77

NADPH 128
Nernst equation 105, 159
– full cell reaction 106

Ohm´s law 77
open circuit potential 101
open circuit voltage 79

overpotential 110, 173
– anodic 110, 173
– cathodic 110, 173
– concentration 111
ozone 125

parabolic trough collector 142
particle ensemble 187
Pauli exclusion principle 73, 188
Peltier effect 86
phase transition 40, 183
photon absorption
– first-law efficiency 153
– second-law efficiency 153
photosynthesis 127
– first-law efficiency 129
– second-law efficiency 129
photosystem II 128
photovoltaic energy conversion 143
– maximal second-law efficiency 154
piston compressor 35
Planck, Max 18, 119
pn-junction 143, 157
power
– final 1
– net 1
– primary 1
power plant
– combined cycle 45, 71
– gas turbine 45, 60
– pumped storage hydro 167, 168
– solarthermal 141
– steam 45
– thermal 45
power to gas 176
pressure ratio 64
– work output optimized 65
process 15
– adiabatic 16, 181
– isentropic 15, 181
– isobaric 15, 179
– isochoric 16, 180
– isothermal 15, 179
– reversible 8, 17, 25
proton exchange membrane (PEM) 116
pumped storage hydro power plants 167, 168

quasi-electrochemical potential 147
quasi-electrochemical potentials 147

radiation losses 131
radiative recombination 149
Rankine cycle 46
– first-law efficiency 49
rate constant 113
reaction
– endothermic 88
– exergonic 90
– exothermic 88
reaction kinetics 112
– losses 111
reaction zone 158
reciprocal space 187
redox potential 101
reflection losses 131
regenerative feed water heating 59
reheat 57
reverse bias 157

Sabatier process 176
saturated vapor curve 184
scattering
– geometric 124
– Mie 124
– Rayleigh 124
second law of thermodynamics 17
Seebeck coefficient 82
– metal 83
– semiconductor 83
selective absorption 126, 135
selective contacts 155
semiconductor 192
– direct 195
– doping 194
– indirect 195
– intrinsic 192
– intrinsic charge carrier density 193
Shockley-Queisser limit 154
short circuit current 159
sign convention 17
solar cell
– a-Si 164
– Auger recombination 162
– c-Si 163
– CIGS 164
– dye-sensitized 165
– e^-/h^+/phonon scattering 162
– impurity scattering 161
– ohmic losses 161

– organic 165
– surface recombination 161
– UI-characteristic 160
solar constant 122
– AM1.5 125
solar irradiation 119
– direct 124
– energy volume density 119
– entropy 123
– exergy 122
– indirect 124
– intensity 119, 121
solarhtermal collector
– first-law efficiency 133
solarthermal collector 131
solarthermal energy conversion 131
– convection losses 131
– heat flow diagram 132
Sommerfeld expansion 191
space charge layer 76
space charge region 76
stack gas release
– first-law efficiency 52, 56
– second-law efficiency 52, 56
standard cell voltage 106
standard hydrogen electrode 101
standard hydrogen scale 100
state variable 15
steady flow equilibrium 34
steam generator 47
– first-law efficiency 51
– second-law efficiency 51, 55
steam mass fraction 185
steam power plant
– first-law efficiency 53, 57
– modifications 57
– second-law efficiency 54, 57
Stefan–Boltzmann constant 122
Stefan–Boltzmann law 122
Stefan-Boltzmann constant 190
stoichiometric coefficient 88
sun/earth system 120
– geometrical factor 121
supercritical region 57, 184
superheat 57
symmetry factor 113

thermal fluid 131
thermalization 145
thermodynamic potential 177
thermodynamic system 15
– closed 15
– isolated 15
– open 15, 31
thermoelectric generator 84
– first-law efficiency 86
thermoelectrics 82
thermovoltage 82
turbine 48
– gas 61
– steam 48
– water 41
– wind 41

valence band 192
van der Waals equation 184
voltage source 78
– electrochemical 106

water splitting 106, 128, 175
– second-law efficiency 175
water turbine 41
wind turbine 41
work 16
– of displacement 20
– flow 31, 33
– technical 32, 33, 35
– useful 19
working fluid 31

www.ingramcontent.com/pod-product-compliance
Lightning Source LLC
Chambersburg PA
CBHW081104220326
41598CB00038B/7219
9781501507632